URBAN HYDROLOGY

Revised Edition

URBAN HYDROLOGY
A Multidisciplinary Perspective

TIMOTHY R. LAZARO

LANCASTER · BASEL

Urban Hydrology—Revised Edition
a **TECHNOMIC**® publication

Published in the Western Hemisphere by
Technomic Publishing Company, Inc.
851 New Holland Avenue
Box 3535
Lancaster, Pennsylvania 17604 U.S.A.

Distributed in the Rest of the World by
Technomic Publishing AG

Copyright © 1990 by Technomic Publishing Company, Inc.
All rights reserved

No part of this publication may be reproduced, stored in a
retrieval system, or transmitted, in any form or by any means,
electronic, mechanical, photocopying, recording, or otherwise,
without the prior written permission of the publisher.

Printed in the United States of America
10 9 8 7 6 5 4 3 2 1

Main entry under title:
  Urban Hydrology: A Multidisciplinary Perspective—Revised Edition

A Technomic Publishing Company book
Bibliography: p.
Includes index p. 239

Library of Congress Card No. 89-51913
ISBN No. 87762-547-6

To Rebecca, Tere and Elena
and
Thanks to Barb Ryan

Table of Contents

Preface xi

Introduction xiii

1. **City Physiology and Anatomy**   1
   1.1 The Rise of Cities 1
   1.2 Forces of Urbanization 3
   1.3 Urban Form Factors 6
   1.4 City Forming Forces 7
   1.5 Summary 10
   1.6 Exercises 11
         References 11
         Selected Readings 12

2. **Impact of Urbanization on Streamflow**   15
   2.1 The Hydrologic Cycle 15
   2.2 Land Use Changes Accompanying Urbanization 19
   2.3 Urban Surfaces and Runoff 19
   2.4 Reduction of Infiltration 23
   2.5 Summary 23
   2.6 Exercises 24
         References 24
         Selected Readings 26

3. **Urbanization and Stream Water Quality**   29
   3.1 Water Quality of Natural Streams 29
   3.2 Impact of Raw Sewage on Stream Water Quality 37
   3.3 Sources of Urban Pollutants 44
   3.4 Impacts of Urban Runoff on Water Quality 47
   3.5 Impacts of Urban Runoff on Stream Water Quality—
         Case Studies 49

	3.6	Impact of Urban Runoff in Erosion	60
	3.7	Impact of Urban Runoff on Stream Water Quality	66
	3.8	Summary	72
	3.9	Exercises	75
		References	75
		Selected Readings	78
4.	**Analysis of Hydrologic Change Due to Urbanization**		**85**
	4.1	How to Start an Urban Hydrologic Study	85
	4.2	Water Quantity Data Collection	88
	4.3	Water Quality Data Collection	91
	4.4	The Urbanized Area and the Watershed	94
	4.5	Estimation of Percent of Imperviousness	98
	4.6	Probabilistic Approaches	101
	4.7	Statistical Techniques	104
	4.8	Urbanized Stream Channels	108
	4.9	Summary	113
	4.10	Exercises	114
		References	114
		Selected Readings	117
5.	**Modeling Urban Water Quantity and Quality**		**121**
	5.1	Mathematical Models	121
	5.2	Modeling a Natural Watershed	125
	5.3	Rainfall-Runoff Modeling	127
	5.4	Urban Watershed Modeling	134
	5.5	Urban Watershed Modeling—Quantity	146
	5.6	Urban Watershed Modeling—Quality	158
	5.7	Summary	169
	5.8	Exercises	171
		References	172
		Selected Readings	176
6.	**Nonstructural Control Measures**		**185**
	6.1	Planning and Planning Commissions	185
	6.2	The Plan and the Planning Process	192
	6.3	Urban Water Resources Planning	201
	6.4	Summary	206
	6.5	Exercises	207
		References	207
		Selected Readings	209
7.	**Structural Control Measures**		**213**
	7.1	Water Quantity	213
	7.2	Water Quality	218

	7.3	Sheet Erosion and Sedimentation Control 223
	7.4	Summary 227
	7.5	Exercises...231
		References..231
		Selected Readings..................................232

**8. Afterword: Perspectives on the Urban
   Hydrologic Problem**   **235**

Index 239

About the Author 243

Preface

JACOB BRONOWSKY, THE superb historian and author of the best seller, *The Ascent of Man*, stated in one of his lectures for the BBC series of the same name (paraphrased):

> . . . all animals leave behind signs of their having been here; only man leaves behind his creations.

Perhaps nothing that man has ever done could be more creative or more indicative of his innovative mind than his design and construction of cities. They are truly one of man's greatest creations, and stand as monuments to his imagination.

Within this immensely powerful intellect lie the solutions to the problems facing cities today. This book considers one of the major problems; that is, the flow of storm water through urban areas to receiving watercourses. Considerable portions of urban budgets are devoted to the safe conveyance of this water, either through sewerage systems or in channels. On an equal par, a related problem concerns the deterioration of the water's quality as it travels through urban areas, reduces the recreation potential of rivers, adds to the cost of water processing, and generates many environmental controversies. All this is occurring within man's creation, even though the most sophisticated technology he has yet devised is available.

I believe that many of the approaches to solving these problems need improvement. The major failing concerns specialization and its spinoff, the lack of effective and earnest communication. It is true that specialization has made our society into a very complex and viable functional unit, but at the same time it has built up formidable communication barriers. An all-too-evident example of this concerns many scientists who, after reviewing a study conducted in a field other than their own — a study written in esoteric terminology — easily overlook the benefits that a full comprehension of this research could provide.

In this manner, much research is duplicated and consequently much time, effort and expense is wasted. Therefore, it is necessary for us to tear down these communications barriers, and cultivate multidisciplinary communication, i.e., to share concepts and knowledge, and ultimately reduce esoteric and costly counterproductive activities. At the onset, it must be acknowledged that, originally, specialization was based on the premise that the specialists were part of a team, and that through teamwork, a sharing and working together, the goal could be more easily accomplished.

I believe the next phase of achievement lies in continuing teamwork and in cultivating our communicative capabilities. Urban water quantity and water quality problems, as well as many other problems, require a multidisciplinary approach.

With this thought in mind, this book is intended to be a multidisciplinary vehicle, to present to scientists in different fields urban hydrology and its problems, and to suggest methods of solution. This book is not intended to be an end in itself, but rather a beginning, a suggestion for other texts that will initiate or contribute to innovative thought by specialists in separate fields, who, by working together, will find answers to critical urban water issues. This book can indeed be a multidisciplinary vehicle, but only if the reader earnestly wishes it to be so. The emphasis is on the communication of concepts, not on definitive solutions. However, it is hoped that the reader will build upon these concepts with an open mind and achieve definitive solutions.

Introduction

> ... it is generally accepted that the contemporary trend toward more intensive urbanization which exists in the United States and in nearly all other nations should continue through the remainder of the century. As a consequence, urban problems associated with the hydrological aspects of water management should become increasingly more acute. The hydrology of urban areas is already quite complex. —Jens and McPherson (1964), p. 20-2.

Urban Hydrology

URBAN HYDROLOGY IS defined as the interdisciplinary science of water and its interrelationships with urban man (Jones, 1971). It is a relatively young science; the bulk of its knowledge has accumulated since the early 1960s. The beginnings of urban hydrology can be traced to the time shortly after the automobile became the major means of transportation in the United States. Roads were paved to facilitate travel, allowing the growth of the suburbs where the commuter escaped the congestion of inner-city life. The result was the rapid creation of large impervious areas, producing noticeable drainage problems. The science of urban hydrology was born out of the necessity to understand and control these problems.

The Need

In 1980, 73.7% of the population of the U.S. lived in urbanized areas (U.S. Department of Commerce, 1986). Each year, urban expansion claims another 420,070 acres (U.S. Department of State, 1971). At the time of this writing, indications are that the 1980 census will clearly demonstrate a continued increase in urban area. The large concentration of people in a relatively small portion of the world's land area has been shown to have a significant influence on environmental processes. McPherson (1974) writes:

> ... the impact of man on the water cycle is greatest per unit area in urban places. Man is capable of transforming his local environment in an almost endless variety of ways, over a matter of a few years, whereas nature moves largely on a timetable of eons. Thus, urban hydrology contends with the dimension of dynamic change because urban development everywhere has been in continuous states of expansion and flux (pp. 15–16).

Many studies have shown that urbanization causes an increase in flooding and drastic changes in water quality. Larger floods increase the risk of property damage and/or injury to residents; the deterioration in water quality causes increases in the costs of water processing (Weibel, 1964) and limits contact recreation along a water course. Flood alleviative measures such as channelization or enlargement of sewers are expensive and, unfortunately, usually represent stop-gap solutions.

In step with the changed hydrology, urban taxes tend to spiral upward. In order to achieve any reduction in these expenditures, appropriate consideration must be given to the physical processes involved. Water has always been, and will continue to be, of prime importance to humankind. There is an increasing need to understand how it is influenced as it flows through the environment where the vast majority of the population continues to congregate.

Scope of This Book

It has been the author's experience in professional planning that there exists a considerable number of communication difficulties between technical, planning and political specialists. Rickert and Hines (1975) observed this phenomenon in water quality studies when they wrote:

> . . . useful assessment of complex river-quality problems requires development of a mutual understanding between the resource scientist and planner. The scientist needs to assess critical problems in the context of a planner's requirements, and the planner needs to appreciate the meaning and implication of the scientists' results. Each must appreciate how the other thinks, and this requires free and continuing communication (p. A4).

The purpose of this book is to outline and discuss the facets of urban hydrology in language which can be easily understood by scholars of different disciplines. In this manner, an appreciation of the roles played by the specialists in various fields is developed. The book supports the view that problems in urban hydrology can only be adequately solved by utilizing a multidisciplinary approach, i.e., that specialists of different disciplines must earnestly work together and seriously attempt to communicate their ideas to one another.

Within the text, urban hydrology is reduced to a simple cause and effect relationship (see Figure I). The first five chapters discuss the cause and effect(s), the sixth chapter presents a brief review of planning problems, and the seventh chapter presents structural means of solution. Brief descriptions of the thrusts of the chapters follow.

1. *The Cause* and its background is discussed in Chapter 1, "City Physiology and Anatomy." Cause is defined as the land use change accompanying ur-

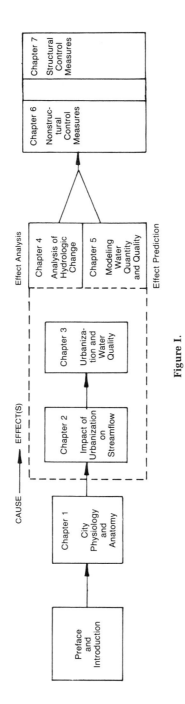

Figure I.

banization, and is discussed in detail in Section 1.2. In order to develop the background for the study of urbanization, the genesis of cities is presented. Then the motivating forces of urbanization are outlined. The last two sections discuss urban morphology. Here, the intention is to give the reader insight into the life processes of a city, so that one might be able to confidently anticipate the physical effects and direction of city growth.

2. *The Effects* on water quantity are discussed in Chapter 2, "Impact of Urbanization on Streamflow," and effects on water quality are addressed in Chapter 3, "Urbanization and Water Quality." These narratives touch on the state-of-the-art understanding of the effects of urbanization on the hydrology and water quality of an area.

3. *The Analysis* of the effects is the concern of Chapter 4, "Analysis of the Hydrologic Change Due to Urbanization." This chapter is a listing with descriptions of the latest methods for identifying changes in water flows and stream morphology caused by the altered streamflow regime following urbanization.

4. *The Prediction* of the effects is discussed in Chapter 5, "Modeling Urban Water Quantity and Quality." The essentials of models are developed, and several types of existing models are outlined. Chapters 2, 3, 4 and 5 bring the reader up to date on the latest developments in urban hydrology.

5. *The Solutions* to the urban hydrologic problem(s) are discussed in Chapters 6, "Nonstructural Control Measures," and 7, "Structural Control Measures." The difficulties that planners encounter are briefly presented and new improved approaches are suggested in Chapter 6. Chapter 7 presents an outline of existing technology which, if employed, could greatly alleviate urban runoff and water quality problems.

This book should be of particular interest to urban water resource engineers and urban water resource planners. Its subject matter is arranged so as to be easily integrated into civil engineering, physical geography, environmental science, or water resource planning curricula. It may also find use as a reference by concerned students and/or professionals.

References

JENS, S. W. and M. B. McPherson. "Hydrology of Urban Areas," Section 20 in *Handbook of Applied Hydrology*. V. T. Chow, ed. New York:McGraw Hill Book Co. (1964).

JONES, D. E. JR. "Where Is Urban Hydrology Practice Today?" *Proc. Am. Soc. Civil Eng., Hydr. Div.*, 97(HY2):257–264 (1971).

MCPHERSON, M. B. *Hydrological Effects of Urbanization*. Paris:UNESCO Press (1974).

RICKERT, D. A. and W. G. Hines. "A Practical Framework for River-Quality Assessment," U.S. Geological Survey Circular No. 715-A (1975).

U.S. DEPARTMENT OF COMMERCE. "Statistical Abstract of the United States," Washington, DC, 985 pp. (1986).

U.S. DEPARTMENT OF STATE. "U.S. National Report on the Human Environment," Department of State Publication No. 8588, Government Printing Office, Washington, DC (1971).

WEIBEL, S. R. et al. "Urban Land Runoff as a Factor in Stream Pollution," *J. Water Poll. Control Fed.*, 36:914–924 (1964).

1 | City Physiology and Anatomy

A COMPREHENSIVE UNDERSTANDING of the physical city (anatomy) and its physiological process of urbanization is essential to the study of urban hydrology. Through such an intimate understanding, future development and decay may accurately be anticipated, and such knowledge can be of invaluable assistance in planning storm sewerage, retention reservoirs and so on. Additionally, if growth can be encouraged which will alleviate urban runoff problems rather than augmenting them, the new construction will complement the surrounding environment and, in the long run, will prove less costly in terms of available natural and fiscal resources. Accordingly, this chapter will briefly present the fundamentals of the physiology and anatomy of cities.

1.1 The Rise of Cities[1]

Why man first settled in communities will forever be shrouded in mystery. At best, scholars can only speculate. One author, Arango (1970), suggests that perhaps one day it became apparent to man that a stationary existence would be far better than constantly following the herds. Gibson (1977), after studying several villages in Mesopotamia, came to the conclusion that the prime reason for settlement was not an economic one, but a result of a preeminently social process. Perhaps the most plausible and most widely accepted reason is the one related to the agricultural revolution.

The agricultural revolution refers to man's successful domestication of animals and grains. This meant that man no longer had to hunt and gather his sustenance, but rather could cultivate crops and practice animal husbandry from a permanent habitation. Boughey (1971) describes these early villages (5000

---

[1]A city is defined by Sjoberg (1965) as ". . . a community of substantial size and population density that shelters a variety of nonagricultural specialists, including a literate elite" (p. 19).

B.C.) as farming communities of about 150 people. Several families worked together to achieve the common goal of cultivating the land, preserving the crops and in general accomplishing all the duties which perpetuated survival.

As technology improved, food surpluses became greater, which permitted communities to develop a more sophisticated social organization and to grow in size. Johnson (1967) states that around 4000 B.C. settlements provided centers for administration and were used not only for the storage of goods, but also for their exchange and distribution. Sjoberg (1965) designates this the preindustrial society, and argues that the food surplus permitted both the specialization of labor and a leadership providing class structure. The preindustrial society had two other important characteristics: writing, which allowed the keeping of records and in general encouraged a cultural tradition and the use of sources of energy other than the muscles of men or animals (i.e., water power, wind).

As the leadership became stronger, at times reinforced by ideology, and as agricultural production became more efficient, cities of considerable size appeared. Uruk and Babylon achieved populations of 50,000 and 80,000, respectively, between 3000 and 2500 B.C. (Haggett, 1973). Communities, however, would remain limited in size by the agricultural productivity of the immediate landscape, and by the limitations of transportation. Johnson pointed out that there were as many as 50–90 agriculturalists in a population to every one non-agriculturalist, indicating that by far the majority of the city dwellers were bound to the land.

Most preindustrial cities maintained ecologically sound relationships. In his description of early Rome, Mumford (1956) defines the relationship between the city and the surrounding landscape as being a symbiotic one. As Rome grew, it began to overtax its environs. At this point, the relationship became parasitic, and Rome was only able to maintain further growth by engaging in a systematic military exploitation of other regions. Rome was not original in this respect; other cities and empires before and after have done much the same. Military might, however, was utilized not to continue the process of urbanization but rather as a stop-gap measure.

Sjoberg suggests that Roman techniques and concepts which were associated with literate traditions diffused into the Arab world and were further developed during the Dark Ages. As Europe came out of the Dark Ages and made contact with the Arab traders, these new ideas had a profound impact on the continent, and also had a direct bearing on the Renaissance.

A thousand years after the fall of Rome, large cities again reappeared. In contrast to Rome and other major powers, they did not depend on the forced appropriation of the natural resources of others, but rather developed friendly trade agreements which all parties found mutually profitable. Mumford designates this the second stage of urbanization.[2] The seventeenth century saw a

---

[2]Urbanization is defined here, after Davis (1965), as the proportion of the total population concentrated in urban settlements, or as a rise in this proportion.

considerable number of cities successfully sustaining a population of about 100,000.

Favorable site location has provided the stimulus for the birth and growth of cities. In their classic paper, Harris and Ullman (1945) discuss three supports for cities:

(1) Cities as central places providing comprehensive services for a surrounding area
(2) Transport cities performing break-of-bulk and allied services along transport routes
(3) Specialized-function cities performing one service such as a trade and social center for a tributary area

If the surrounding region is flat and of equal productivity as in the American middle west, these centers tend to be uniformly spaced and the communities fall into a hierarchy of sizes according to the range of functions they provide.

Break-of-bulk refers to changing from one form of transportation to another, for example, from ship to train. Whenever this process occurs, a potential exists for the manufacturing of the goods received and the development of value-added services. Essentially, several of the urban-forming forces become available, and a city can easily evolve. Many cities such as New York, Louisville, San Francisco and Chicago started as break-of-bulk points. In recent years, truck hauling and the construction of arterial highways have caused many cities to grow because they happened to be at the junction of two interstate highways.

Specialized-function cities are supported by a highly localized resource. Miami is an example; the climate and the beaches have created a splendid milieu for relaxation and recreation. Scranton and Wilkes-Barre, Pennsylvania, are specialized coal mining centers. These cites have acted as nuclei for related functions, and many communities in the surrounding areas have grown in response. Harris and Ullman (1945) go on to say that most American cities are an integration of the three types discussed previously.

1.2 Forces of Urbanization[3]

Around the year 1800, a trend related to city development was observed. Rugg (1972) points out:

> ... beginning with the nineteenth century, a new trend becomes visible—that of "urbanization." Up to this time, we have witnessed what might be called the development of the "city," a form of settlement in which only a small proportion of the total population lived (p. 49).

---

[3]Urbanization is defined (combining Davis, 1965 and Savini and Kammerer, 1961) as the concentration of people in urban settlements and the process of change in land use occupancy resulting from the conversion of rural lands into urban, suburban and industrial communities.

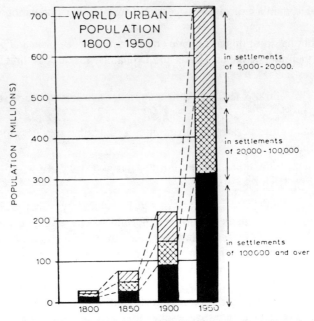

Figure 1.1. World urban growth [*Source*: Johnson (1967)].

Rugg goes on to note that after 1800, cities began to accumulate greater and greater proportions of the population, and, as time went by, this urbanizing trend accelerated.

Johnson (1967) shows that from 1800 to 1950, world urban population grew from less than 40 million people to greater than 700 million: an increase of about 1650% (Figure 1.1). Hoyt (1963) estimates that in 1800, 1% of the world's population lived in cities of 100,000 or more. In 1960, he estimated that the proportion had increased to 20%. Rugg (1972) demonstrates that 36 of the major cities of the world experienced an average population increase of about 200% from 1800 to 1850, and about 200% again from 1850 to 1890. For example, London and New York, the two largest cities, went from 958,800 and 62,900, respectively, in 1800 to 4.2 million and 2.7 million in 1890. Bogue (1953) points out that standard metropolitan areas[4] within the United States had grown 177.8% in the period from 1900 to 1950. Clawson and Hall (1973) describe urbanization since the second world war as "suburbanization," and note that much of rural America tended to lose population in the 1940s, 1950s and 1960s. Muller (1976) states that 37.6% of the U.S. population lived in the suburbs in 1970; in 1975, it is estimated that 39.1% were suburban dwellers.

One is left wondering what could be the cause of such a vast relocation in

---

[4]Except in New England, a "standard metropolitan area" is a county or group of contiguous counties which contains at least one city of 50,000 inhabitants or more (Bogue, 1953).

world population. This section will discuss some of these forces that led to this logarithmic impact.

Revival of thinking in the post-Renaissance era gave impetus to the acceptance of science and scientific methods.[5] The procedure of developing a hypothesis, testing it, observing the results and seeking a scientific solution to problems was clearly formulated. All scientific disciplines were to benefit by advances at these fundamental levels. Ultimately, technology was to make quantum steps forward. Through the use of the latest scientific methods and technology, machines were designed and built stronger and faster than any man could hope to be. Coupling these new-found abilities with the profit incentive provided by Keynesian economics, we begin to see the rudiments of the industrial revolution. Technology was applied to make a profit, and the quest for improvements in technology increased the profit.

Advances in mechanization allowed unprecedented growth in agricultural production efficiencies and manufacturing. Johnson points out that there was a change from subsistence agriculture to an agriculture based entirely on production for sale. Agriculture itself became a cost-effective industry, supplying goods to urban areas and adopting scientific methods to improve and expand its own production processes.

As agricultural endeavors became more efficient, millions of people were freed from the constant toil of cultivation and were attracted by hopes of lucrative employment in the cities. The labor force was available, and large scale industrialization was under way. Industrial development is closely linked to cities and urbanization.[6] Clawson and Hall (1973) point out that industries could often best develop in cities, since there was an available labor supply and an access to raw materials and markets. In response to this, factories either located in established cities or provided a catalyst for the creation of cities. For every person employed in the basic sector of employment (i.e., factories), three or four had to be employed in the nonbasic (or service-oriented) category (Alexander, 1954). The growth of service-oriented businesses (i.e., barbers, laundries, shoemakers) accompanied the expansion of industries, multiplied city employment opportunity; as a result, population and cities grew phenomenally.

Davis (1965) notes that industrialization had a strong influence on increasing the number of people living in English and Welsh cities. He points out that by 1801 about a tenth of the population in these countries was living in cities of 100,000 or larger[7]; in 40 years this proportion doubled, and in another 60 years it doubled again. Davis goes on to illustrate that there is a close relationship between economic development and urbanization.

---

[5] See Bronowski (1973) for an excellent discussion of this period.

[6] Davis (1965) points out that city growth and urbanization are not necessarily the same process.

[7] Davis uses the proportion of people living in cities of 100,000 or more as an index to urbanization.

Further advances in science and technology within the 19th century not only enhanced the urban living, but further accelerated city growth. Improvements were made in transportation as the railway and steamship were invented. Movement between cities as well as within cities was facilitated in this manner, and large scale trade was established on a worldwide basis. Cities became the centers of communication, and new ideas were brought by immigrants from distant lands. These people arrived with a willingness to work, and augmented the labor force.

Advances in technology continued to permeate all disciplines, and the results were remarkable improvements in industrial processes. The concept of mass production (of automobiles) was borrowed and used in other fields, e.g, the mass production of houses. Mass production made the individual worker into a specialist. The concept of specialization not only influenced industrial organization, but affected research, and through specialization technology improved further. Today, this upward spiral continues. Urbanization has been and is still an integral part of the American epoch, as well as of other industrialized countries of the world, and its close affinity with industrial methods rapidly promoted metropolitan growth.

The forces of urbanization are a product of man's genius, of his continuous quest for efficiency and of his need for the social and cultural milieu that an urban area can provide. His continuously improving technology has given him the power to realize his abstract ideas. Within cites, he has found the heterogeneity of human ideas in which to expand. Cities are a product of man's imagination and man's thinking mind. As to the future of urbanization, Gibson (1977) perhaps says it best when he states: ". . . the citizens of the city will command its future, as they probably always have, though they have not always realized it" (p. 14).

1.3 Urban Form Factors

Up to this point, we have discussed in general terms the physiology of cities. In this section, we will examine the factors which direct anatomical growth. The concepts of physiology and anatomy of cities are interrelated, and cannot totally be separated. The three factors which have directed urban form are: (1) topographic obstacles, (2) geology and (3) the street grid. Topographic obstacles may be mountains, water bodies, large rock outcroppings, swamps, etc. A city can only spread up to one of these obstacles, and then it will begin growth in a direction of less resistance. Numerous cities display the results of this type of growth. In some cases, cities have grown vertically because of several area limitations imposed by topographic constraints.

Geology of the city area can limit the capability of the region to support large structures. Legget (1973) points out that the solid Manhattan schist underlying New York City provided a firm foundation for the erection of skyscrapers; on the other hand, London clay was not amenable to skyscrapers, and consequently, vertical growth of London was limited until recently.

By far the major controlling influence on city form has been the street grid pattern. Martin (1972) points to this factor when he states:

> ... the grid of streets and plots from which a city is composed is like a net placed or thrown upon the ground. *This might be called the framework of urbanization.* That framework remains the controlling factor of the way we build, whether it is artificial, regular and preconceived, or organic and distorted by historical accident or accretion (p. 10).

The majority of the world's cities have adopted the grid pattern, and their form has been limited by it. Width of blocks and streets have a profound impact upon urban form. LeCorbusier (1939) points out that by increasing the size of the street net in Manhattan and using a different building form, skyscrapers could be eliminated, and a city having most of its area in parks would be created. This type of city design will be described further in another part of this book, since the adoption of this design would have a significant influence on urban hydrology.

It must be noted that these three limiting factors are in turn limited by the capabilities of the available technology and the preceding urban growth. In other words, growth in any one period will tend to occur on lands which are readily available, even though growth in the past might have produced structures which were not so efficient in the utilization of space. Following along similar lines, cultural limitations may have a similar influence; zoning and growth management ordinances may direct growth and influence urban anatomy.

Urban anatomy is also affected by external factors such as the capabilities of the transportation media and the degree of trade relationship with the surrounding region and other areas.

In summary, urban form is a result of environmental and cultural factors and the historical period and its technological capabilities.

1.4 City Forming Forces

In this section, we will outline some city physiological factors and discuss their impact upon urban form.

The form and arrangement of the early communities (preindustrialization) were dictated by human needs. Boal (1968) refers to these urban areas as "pedestrian cities." As the name implies, movement by foot was the factor determining the size of the early city. Buildings of commercial and administrative significance were situated close together and roads were narrow, not much wider than horse trails. The upper classes demanded easy access to the city center, and therefore lived in the downtown areas with the lower classes being pushed outward to the edges. During this period, urban growth occurred by building new cities and not by adding to existing ones.

The exceptions to this were found in the capital cities. Many were contemporary architectural masterpieces, such as Rome, Alexandria, Athens, etc., and were built to glorify (and at times to deify) the reigning authority, and not purposefully constructed in response to human needs.

The steam engine and wheel track city came into being after the middle of the 19th century, as the tracks were driven outward from the center of the city. Urbanization followed along the rails of the new tranportation medium with the greatest population concentrations around the stations, thus creating a star-shaped city.

By and large, the cities of today may be classified as "flexible cities" (Boal, 1968). This type of city grew in response to the spread of electricity, the development of the internal combustion engine, telecommunications and the steel frame building. It must be noted that many flexible cities spread around a core which was the older, wheel track city. The widespread distribution of electricity made possible the addition of mass transportation to many new routes within the urban area, and permitted access to and consequent urban development in between the points of the older, star-shaped city.

The invention of and improvements in telecommunications and the steel frame building had profound influences upon industrialization. The telephone made it possible for executive decisions to reach the most remote parts of the factory. Rapid communication with the outside kept the industry abreast of any changes in the external market. The latest information was quickly obtainable and the world shrunk as telecommunications were improved.

The steel frame building, by adding more floor space to the downtown area, changed the morphology of the community. Cities expanded vertically. Many manufacturing processes could be enhanced by a vertical orientation. Other processes adapted to it. Johnson (1956) observed that businesses which were functionally related could congregate in the same building, creating an efficient symbiosis. Apartment complexes were constructed, housing thousands of people and increasing population densities. The large urban-suburban commercial-industrial and residential complexes of today would not have been possible without the development and successful employment of the steel frame building.

The automobile stands as the greatest instrument of change in urban and suburban form. The worker was liberated from having to live in proximity to his place of employment, and commuting on a daily basis to the newly expanded "suburbs" became a normal part of life. Residential satellites, chiefly sleeping places, as Murphy (1966) notes, sprang up in response to the escape to the suburbs. Soon, these satellites grew and coalesced, developing into cities by themselves.

The results of these urban forming forces have been classified by Harris and Ullman (1945) into three generalizations for the internal structure of cities: (1) concentric zones, (2) sector theory, and (3) multiple nuclei (Figure 1.2). The concentric zone theory (Burgess 1925) postulates that a city may be divided into five concentric rings. Progressing from the center outward, these are (1) the central business district, which is the focus of social, commercial and civic life; (2) the zone of transition, which encircles the downtown area and is an area of residential deterioration in many cities, inhabited by recent immigrants; (3) the zone of independent workingmen's homes, inhabited by workers who have escaped the city and still desire to live within easy access of their work; (4) the zone of better residences, made up largely of single family homes and high-class apartments; and (5) the commuter's zone, the suburbs or satellite city.

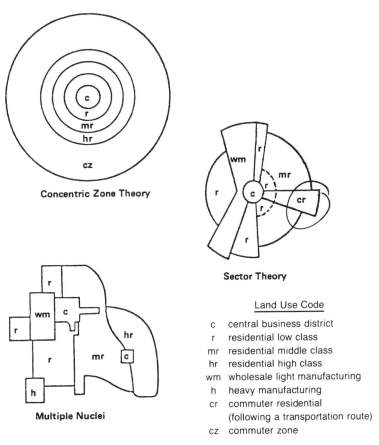

Figure 1.2. Generalizations of the internal structure of cities. Modified from Harris and Ullman (1945) (p. 5).

The sector theory (Hoyt, 1939) states that urban growth occurs along the main transporation routes to form the star-shaped city. The city is still considered a circle, and has various sectors radiating outward from the center. Similar types of land use begin at the center and follow outward to the edge.

The multiple nuclei theory suggests that the land use pattern is not built around a single center, but around several discrete nuclei. These nuclei might have always been present, but not very pronounced, or they may have developed as urban-forming forces stimulated by specialization and/or migration.

These three generalizations have been cause for much debate. Few cities exist which may serve as clear examples of any one pattern. Most cities have parts of all three theories within their spatial structures.

1.5 Summary

Cities have developed in response to human social and economic needs. The first communities were based on agriculture and consisted of several families joining together in a joint venture of cultivation. All mature and healthy members of the society worked for community survival. As time passed and technology improved, a few men were freed from agricultural toil and were able to pursue other occupations, and classes began to develop.

These early cities were located in agriculturally rich regions and grew to a size which was a function of the needs of the people, the productivity of the surrounding landscape and the capabilities of the transportation media.

As technology improved, more people were freed from agriculture, and cities began to develop which were located to satisfy particular functions. For example, a seaport city on a natural bay provided the service of unloading ships, packaging the cargo and shipping it by camels to other points.

The revival of thinking generated by the Renaissance had a profound influence on technology and the acceptance of scientific methods which together enabled men to build and maintain large cities.

Around 1800, a new stage of city building began; the process of urbanization became a reality. Working hand in hand with urbanization, the processes of industrialization affected not only city building practices, but also rural cultivation methods. In this manner, significant improvements in agricultural production efficiencies were brought about, and workers freed from the farms flocked to jobs in the cities. Ultimately, more people lived in cities than in rural areas. Urban areas grew phenomenally in population and area in response to the population shift. Urban form reflected the capabilities of the technology through the transporation media, construction technology and the physical constraints imposed by terrain.

In the early twentieth century, the invention of the automobile radically changed urban form by enhancing daily commuting. Suburbs developed which grew by themselves into small cities, and by coalescing enlarged the overall urbanized area. The end result was the creation of the huge metropolitan complexes of today.

As to the future of cities and urbanization, it appears that the growth trend will continue, but at a slower rate.

1.6 Exercises

Considering an urban area with which you are very familiar:

1 Could this area be modeled after one of the models in Figure 1.2?
2 How does it differ from these models? Was it the topography or the history of the area that created differences?

References

ALEXANDER, J. W. "The Basic-Nonbasic Concept of Urban Economic Functions," *Econ. Geo.*, 30(3):246–261 (1954).

ARANGO, J. *The Urbanization of the Earth*. Boston:Beacon Press, 175 pp. (1970).

BOAL, F. W. "Technology and Urban Form," *J. Geog.*, 67(3):220–236 (1968).

BOGUE, D. J. "Population Growth in Standard Metropolitan Areas, 1900–1950," Housing and Home Finance Agency, Washington, DC (1953).

BOUGHEY, A. S. *Man and the Environment*. New York:Macmillan Publishing Co., 472 pp. (1971).

BRONOWSKI, J. *The Ascent of Man*. Boston:Little, Brown and Company, 448 pp. (1973).

BURGESS, E. W. "The Growth of the City: An Introduction to a Research Project," in *The City*. Park, Burgess and McKenzie, eds. Chicago:University of Chicago Press, 239 pp. (1925).

CLAWSON, M. and P. Hall. *Planning and Urban Growth: An Anglo-American Comparison*. Baltimore:Johns Hopkins University Press, 300 pp. (1975).

DAVIS, K. "The Urbanization of the Human Population," *Scientific Am.*, 213(3):40–54 (1965).

GIBSON, J. E. *Designing the New City: A Systematic Approach*. New York:John Wiley and Sons, Inc., 288 pp. (1977).

HAGGETT, P. *Geography: A Modern Synthesis*. New York:Harper and Row, Publishers, Inc., 483 pp. (1972).

HARRIS, C. D. and E. L. Ullman. "The Nature of Cities," *Ann. Am. Acad. Pol. Soc. Sci.*, 242:7–17 (1945).

HOYT, H. "The Structure and Growth of Residential Neighborhoods," Federal Housing Administration, Washington, DC (1939).

JOHNSON, J. H. "The Geography of the Skyscraper," *J. Geog.*, 4(8):379–387 (1956).

JOHNSON, J. H. *Urban Geography*. Oxford:Pergamon Press, 183 pp. (1967).

LE CORBUSIER. *Oeuvres Completes 1934–1938*. Zurich, Switzerland:Girsbinger (1939).

LEGGETT, R. F. *Cities and Geology*. New York:McGraw-Hill Book Co., Inc., 181 pp. (1973).

MARTIN, L. "The Grid as Generator," in *Urban Space and Structures*. L. Martin and L. March, eds. Cambridge:Cambridge University Press, Chapter 1 (1972).

MULLER, P. O. "The Outer City: Geographical Consequences of the Urbanization of the Suburbs," American Association of Geographers, Resource Paper No. 75-2, Washington, DC (1976).

MUMFORD, L. "The Natural History of Urbanization," in *Man's Role in Changing the Face of the Earth*. W. L. Thomas, ed. Chicago:University of Chicago Press, pp. 382-398 (1956).

MURPHY, R. E. *The American City: An Urban Geography*. New York:McGraw-Hill Book Co., Inc., 464 pp. (1966).

RUGG, D. S. *Spatial Foundations of Urbanism*. Dubuque, IA:William C. Brown Co., 313 pp. (1972).

SAVINI, J. and J. C. Kammerer. "Urban Growth and the Water Regimen," U.S. Geological Survey Water Supply Paper No. 1591-A, 43 pp. (1961).

SJOBERG, G. "The Origin and Evolution of Cities," in *Cities: Their Origin, Growth and Human Impact, Readings from Scientific American*. pp. 19-27 (1965).

Selected Readings

Berry, B. F. L. *The Human Consequences of Urbanization*. New York:St. Martin's Press, 205 pp. (1973).

Carter, H. *The Study of Urban Geography*. London:Edward Arnold Publishers, 346 pp. (1972).

Clawson, M. *Suburban Land Conversion in the United States: An Economic and Governmental Process*. Baltimore:Johns Hopkins University, 406 pp. (1971).

Davis, K. "Cities: Their Origin, Growth and Human Impact," in *Readings from Scientific American*. San Francisco:W. H. Freeman & Co., 297 pp. (1973).

Fleisher, A. "The Influence of Technology on Urban Forms," *Daedalus*, 90:48-60 (1961).

Forrester, J. W. *Urban Dynamics*. Cambridge, MA:The MIT Press, 285 pp. (1969).

Frey, H. T. "Major Uses of Land in the United States, Summary for 1969," USDA, Agricultural Economics Report No. 247 (1973).

Gallion, A. B. *The Urban Pattern*. New York:D. Van Nostrand Co., 446 pp. (1950).

Helmer, O. et al. "Technology and the Study of the City," Task Force on Economic Growth and Opportunity, Chamber of Commerce of the U.S., Washington, DC, 50 pp. (1968).

Hines, F. K. et al. "Social and Economic Characteristics of the Population in Metro and Nonmetro Counties, 1970," USDA, Economic Research Service, Agricultural Economic Report No. 272 (1975).

Horton, Frank E. et al. "Urban Growth and Development Models: Transition and Prospect," *J. Geog.*, 70(2):73-83 (1971).

House, P. *The Urban Environmental System*. Beverly Hills, CA:Sage Publications, 316 pp. (1973).

Hoyt, H. "The Growth of Cities from 1800 to 1960, and Forecasts to the Year 2000," *Land Econ.*, 39(2) (1963).

Le Corbusier. *When the Cathedrals Were White*. New York:Reynolds and Hitchcock (1947).

McHarg, I. L. *Design with Nature*. Garden City, NY:The Natural History Press, 198 pp. (1969).

Meyerson, M. D. and R. B. Mitchell. "Changing City Patterns," *An. Am. Acad. of Pol. Soc. Sci.*, 242:149-161 (1945).

Morris, A. E. J. *History of Urban Form*. London:George Godwin, Ltd., 268 pp. (1972).

Multnomah County Planning Commission. "Urban Forms: An Introduction to the Concept and a Review of Some Factors Which Influence the Shape of the Community," 72 pp.

Ottensmann, J. R. *The Changing Spatial Structure of American Cities.* Lexington, MA:D. C. Heath and Co., 207 pp. (1975).

Stanislawski, D. "The Origin and Spread of the Grid Pattern Town," *Geog. Rev.*, 36:105–120 (1946).

Sweet, D. C., ed. *Models of Urban Structure.* Lexington, MA:D. C. Heath and Co., 252 pp. (1972).

Wakstein, A. M. *The Urbanization of America.* Boston:Houghton Mifflin Company, 502 pp. (1970).

Winsborough, H. W. "City Growth and City Structure," *J. Reg. Sci.*, 4:35–49 (1962).

Yuill, R. S. "A General Model of Urban Growth: A Spatial Simulation," Michigan Geography Publication No. 2, Department of Geography, University of Michigan, Ann Arbor, MI, 221 pp. (1970).

2 | Impact of Urbanization on Streamflow

> . . . of all the land use changes affecting the hydrology of an area, urbanization is by far the most forceful. —L. B. Leopold (1968), p. 1.

CHAPTER 1 PRESENTED the causes of urbanization. This chapter will investigate urban land use, land surface changes and their effect on the hydrologic cycle.

2.1 The Hydrologic Cycle

The volume of global water is a constant; Peixoto and Kettani (1973) estimate that the oceans contain 97.2% (1.35 × 10^{18} m^3) of water, glaciers and polar ice 2.09%, aquifers 0.6%, lakes and rivers 0.014%, and the atmosphere a relatively miniscule amount. Barry (1969a) points out that at any instant, rivers and lakes hold only 0.33% of all freshwater and the atmosphere a mere 0.035%. The continuous motion of this water is explained by the hydrologic cycle. A natural machine powered by the sun, the cycle removes water from the oceans into the atmosphere and onto the land, where it is pulled by gravity and flows back to the oceans. Two branches make up the cycle: the atmosphere, in which the horizontal movement of water is mainly through the gaseous phase, and the terrestrial branch, in which the flow of water is mainly by liquid, although the solid phase predominates.

The hydrologic cycle begins when moisture is evaporated from the ocean. Within the atmosphere, the water vapor condenses into clouds and precipitates (Figure 2.1) onto the land surface, reaching the vegetation first. Portions of these waters may be evaporated and reenter the atmospheric branch, or become surface or subsurface runoff. Surface runoff is temporarily detained as it fills depressions, and then any excess flows into streams. Subsurface runoff penetrates the soil and moves slowly, seeping into streams. Plants may withdraw moisture from the subsurface runoff for their physiological needs, and through their stomata moisture is transpired into the atmosphere.

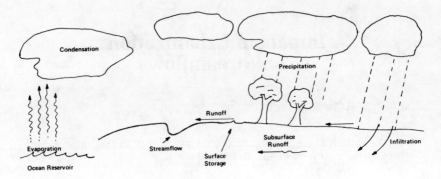

Figure 2.1. The hydrologic cycle.

EVAPORATION

Barry (1969b) discusses the mechanisms of evaporation. One gram of water requires 540 cal of heat at 100°C and 600 cal at 0°C in order to vaporize. Dry winds as well as direct sunlight enhance the evaporation process.

PRECIPITATION

Water vapor condenses around minute dust particles found in the atmosphere. At first, the moisture forms very fine droplets, less than 0.04 mm in diameter. Davis and DeWiest (1966) suggest that clouds are composed of colloidal suspensions of these small particles. For detailed discussions of this phenomenon, see Flohn (1969), Viessman et al. (1977) and, for the measurement of precipitation, Gilman (1964) and Singh and Buapeng (1977).

INFILTRATION

Rain falling on a natural area of forests and meadows is intercepted by the leaves and branches of trees and smaller plants. When the vegetation cover becomes saturated, the subsequent rainwater begins to drip onto the ground, penetrating the soil. Kirkby (1969) stated, "Infiltration rate is defined as the maximum rate at which water can penetrate into the soil" (p. 215). Soil porosity and antecedent soil moisture conditions, i.e., how wet or dry the soil is prior to the rainfall, are perhaps the two most important variables affecting infiltration. Philip (1957) expressed the components of infiltration in the equation:

$$i = St^{1/2} + At$$

where

i = the cumulative infiltration
S = the sorptivity
t = the time
A = second parameter, which is also related to the analysis developed by Philip

SURFACE STORAGE

Vegetation as well as small depressions may detain significant quantities of water, actually storing the moisture, delaying its flow and allowing some rainwater to reevaporate into the atmosphere.

RUNOFF

Davis and DeWiest (1966) define runoff as streamflow, i.e., the sum of surface runoff and subsurface (groundwater flow) runoff. Surface runoff equals precipitation minus the sum of surface storage and infiltration. In other words, when the surface storage and the soil become saturated, infiltration ceases and subsequent rainfall becomes surface runoff. Subsurface runoff is rainwater which infiltrates the surface and flows much more slowly on its way to a stream than does surface runoff.

STREAMFLOW

Streamflow is composed of surface and subsurface runoff, and the small amount coming from rain falling directly into the stream. Streamflow is termed "discharge" and is computed by the formula:

$$Q = VA$$

where

Q = discharge in ft^3/sec or m^3/sec
V = velocity in ft/sec or m/sec
A = area in ft^2 or m^2

HYDROGRAPH

Streamflow changes following a rainstorm may be analyzed by means of plotting discharge against time. Figure 2.2 represents such a plot of points and

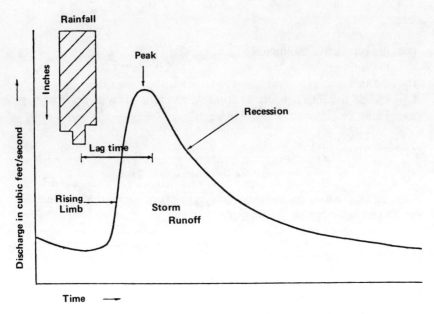

Figure 2.2. Example of a hydrograph. Modified from Chorley (1969).

is referred to as a "hydrograph." Linsley, Kohler and Paulus (1949) divided the hydrograph into three parts: (1) the rising limb, (2) the crest segment, and (3) the falling limb, or recession.

Hydrograph shape varies with the magnitude and the spatial and temporal distribution of precipitation, singularly or in concert. These factors may cause sharp peaks with fast-rising limbs, double peaks, broadened peaks, etc. As a rule, one can assume that no two hydrographs are alike and the variety for a single watershed is indeed infinite. In an attempt to arrive at a basis for comparison of hydrographs, a standardized method was proposed by Sherman (1932) and called the "unit graph." Chow (1964) and Nash (1966) have presented techniques which may be used to derive the "unit hydrograph" (an extension of Sherman's idea). The term "unit" refers to a hypothetical rainfall of 1 cm or 1 in. distributed uniformly in space and time over the catchment.[8]

Unit hydrographs have become powerful analytical tools employed by hydrologists in determining watershed characteristics, precipitation effects on the watershed and the influence of land use changes on the discharge characteristics of a stream. In general, if one of the land-phases of the hydrologic cycle is varied with the others held constant, hydrograph shape will also vary. If there is a decrease in the quantity of flow-retardant vegetation in a watershed, the storage factor will be decreased, runoff will increase, and the unit hydro-

---

[8]The unit hydrograph will be discussed in greater detail in Chapter 5.

graph will rise higher. The rising limb of the hydrograph is usually smoothly sloped, because rainwater is delayed by its accumulation in storage before excess becomes available for runoff. In the modified watershed, the rise (lag) time is reduced, and the hydrograph peaks higher.

2.2 Land Use Changes Accompanying Urbanization

Stankowski (1972) described these changes quite accurately when he observed that:

> ... urbanization begins with the occupancy of rural lands by small, concentrated communities with close groupings of homes, schools, churches and commercial facilities. Further growth is characterized by large residential subdivisions, additional schools, shopping centers, some industrial buildings, and an enlarged network of streets and sidewalks. Central business districts evolve; these contain large stores and offices and often cultural and civic centers. Industrial growth continues along waterways, railroad lines, and major highways. The process continues until homes, apartment complexes, commercial and industrial buildings, streets, parking lots, and sidewalks occupy all or most of the former rural land area (p. B220).

In an attempt to classify these land use changes, Savini and Kammerer (1961) divided urbanization into four stages: (1) rural, (2) early urban, (3) middle urban, and (4) late urban. The rural stage is when the area under study is in the virgin stage, under cultivation or in pasture. Quite obviously, most of the earth's land surface is in this stage.

Early urban land use is characterized by city-type homes built on large plots with much of the indigenous vegetation remaining intact. Many small rural communities and suburban areas may be so classified.

The middle urban stage concerns the construction and growth of large-scale housing developments, shopping centers, schools, churches and industries, all of which are attended by more and more areas devoted to streets and sidewalks. The middle urban stage of development is found mainly in suburban areas.

The late urban stage is brought about as a result of even more development, which may cause the remaining amount of indigenous vegetation to diminish to zero and the land surface to approach a total cover of manmade structures and accoutrements.

2.3 Urban Surfaces and Runoff

As the land surface is developed for urban use, a region is transformed from the natural state to a totally manmade state. New structures add large amounts

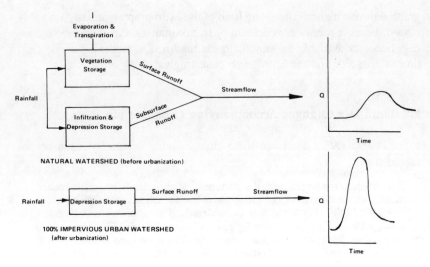

Figure 2.3. Comparison of natural and urban watersheds.

of impervious areas[9] to the watershed, which in general increase slopes and considerably diminish the water storage capability. As the area covered by structures approaches 100%, the amount of vegetation, natural surface and infiltration will all approach zero.

Figure 2.3 schematically illustrates the two extremes: a natural watershed and a totally urbanized watershed. Part of the rain falling into the natural watershed is intercepted by the vegetation, and the remainder falls on the ground, is stored in depressions and begins to penetrate the soil. As the vegetation and the soil become saturated, the excess rainfall begins to run off the surface; subsurface flow has also begun. However, since the subsurface flow is slower than surface runoff, it will take longer before the subsurface flow contributes to the streamflow.

Saturation and consequent surface runoff occur relatively rapidly in the urban watershed, since storage and infiltration have been reduced to practically zero. Incoming rainwater quickly fills any depressions and becomes readily available for surface runoff. Subsurface runoff is virtually nonexistent, and most excess rainfall augments streamflow.

The unit hydrographs (Figure 2.4) demonstrate the results of similar storms occurring over a watershed which was progressively experiencing urbanization. Because storage is diminishing as the watershed becomes urbanized, a larger volume is more readily available for runoff and the hydrographs rise more abruptly, attaining higher peaks as imperviousness increases. The falling

---

[9]Impervious area is defined after Fleming (1975) as ". . . the land surface with an infiltration capacity equal to zero" (p. 107).

limb follows the same pattern, i.e., it is more steeply sloped as it returns to baseflow. Since recessions for the earlier storms were augmented by subsurface runoff, the hydrographs took considerably longer to return to baseflow.

Several researchers have demonstrated the effects of increased imperviousness on the flood hydrograph, employing laboratory catchment models. Roberts and Klingeman (1970), in one of their experiments, tested the effects of a 0%, 50% and 100% permeable surface on the unit hydrograph. It was found that the rising limbs on the 0% and 50% permeable surfaces were coincident for approximately half of their lengths, at which point the 50% permeable surface lagged the impervious one (Figure 2.5). This phenomenon occurred because the lower half of the basin was impervious and the water was running off this portion at a volume and a rate which was sufficient to preliminarily follow the 0% permeable hydrograph. However, water was infiltrating the upper half of the 50% permeable basin, and this water volume was not readily available to complement the water flowing off the lower portion, i.e., a lagtime was created. The result was that the 50% permeable hydrograph separated from the impervious one and began to slope more smoothly. On the other hand, the 100% permeable watershed was observed to produce a significantly slower risetime (about twice that of the 50% permeable hydrograph) demonstrating the effects of the time lag caused by greater storage and infiltration capabilities. Other model results obtained by Black (1970) essentially confirm those obtained by Roberts and Klingeman.

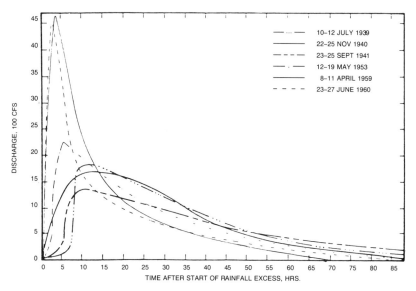

Figure 2.4. Unit hydrographs [*Source*: Moore and Morgan (1969); © 1969 by the University of Texas Press (p. 217)].

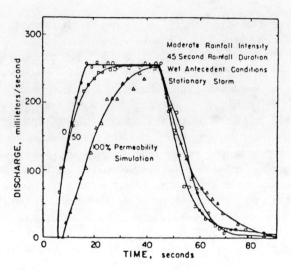

Figure 2.5. A unit hydrograph comparing surfaces of 0, 50 and 100% permeability [*Source*: Roberts and Klingeman (1970) (p. 405)].

Similar experiments cannot be easily conducted outside the laboratory since there are a multitude of climatic and physiographic variables to be considered. However, many studies have demonstrated, and by now it is well-known, that impervious areas do indeed affect the volume and timing of surface runoff. Harris and Rantz (1964) studied the impact of urban growth on a 5.12-mi^2 watershed and observed that storm runoff had increased substantially. Crippen (1965) employed unit hydrograph methods on the Sharon Creek watershed, a small (245-ac) basin, and found that peak discharge (Q_p) increased from 180 cubic feet/second (cfs) to 250 cfs as urbanization proceeded over the several-year study period.

Espey, Winslow and Morgan (1969) studied several previous works completed on peak floods for urban areas. Two of their conclusions may be taken as a reasonable estimate of what may be expected to occur to the unit hydrograph as a region undergoes urbanization: (1) unit hydrograph peak flows may be approximately tripled, and risetime reduced by a factor of approximately one-third and (2) the peak storm runoff may be expected to be from two to four times that of the undeveloped area, depending upon channel modifications, impervious cover, amounts of vegetation in channel and drainage facilities.

Drainage in most urban areas is facilitated by sewers. In his study of portions of northern Virginia, Anderson (1970) reported:

> . . . improvements of the drainage system may reduce lagtime to one-eighth that of natural channels. This lagtime reduction, combined with an increased storm runoff resulting from impervious surfaces, increases the flood peaks by a factor that ranges from two to nearly eight (p. C-1).

Brater and Sangal (1969) discussed the sewerage design[10] problems created by the need to convey the greater volumes and peaks produced by urbanization.

2.4 Reduction of Infiltration

Thus far, we have considered the effects of urban land use on surface runoff. Since streamflow is affected in a dramatic manner, these effects may be clearly illustrated by observing changes in the unit hydrograph. It has been shown that as an area becomes impervious, infiltration is reduced and subsurface runoff diminishes. Reduction in subsurface flow has a far less dramatic impact than surface runoff, since it is not a demonstrative portion of the unit hydrograph. There are two basic reasons for this: (1) subsurface runoff may take days to reach a stream, and therefore may not appear as part of the unit hydrograph, since the hydrograph considers only the relatively short storm period and (2) depending upon the time of year, vegetation may absorb significant quantities of subsurface runoff and evapotranspire moisture into the atmosphere. Sawyer (1963) studied annual runoff[11] from two basins on Long Island. One was affected by urbanization, the other unaffected. He estimated that the affected stream was losing 2% of its subsurface runoff to surface runoff. Few studies have been published concerning this water transfer; however, it would be quite accurate to estimate that this phenomenon is indeed occurring in urban basins. The volume of water lost by an urban basin to surface runoff would directly depend upon its percent of imperviousness.

Quantity and species of vegetation have been studied by watershed managers to define potential water yield. Technical Report NE-13 (1975) by the U.S. Department of Agriculture, Forest Service, presents papers by managers discussing the various aspects of forest maintenance and its anticipated effects on water yields. Hewlett and Helvey (1970) reported that stormflow volume was significantly (0.001 level) increased by 11% overall after clearfelling a 108-ac forest catchment in the southern Appalachians. This may affect the streamflow regime (peaks and baseflow) in a similar but much less dramatic manner than urbanization.

2.5 Summary

The hydrologic cycle considers the continuous circulation of water from the ocean reservoirs into the atmosphere, onto the land and eventually back to the

---

[10]Sewerage design will be briefly presented in Section 5.4.

[11]"Annual runoff," for the purposes of this book, is defined as the depth to which the drainage area would be covered if all the runoff for a year were distributed uniformly over its surface.

oceans. Urbanization brings with it changes in land use, at times severely disrupting the natural landscape, replacing it with impervious surfaces and redistributing the land surface flows of the hydrologic cycle. Storm water runs off these modified surfaces more rapidly, and in many cases drainage is facilitated by sewerage. The results may be demonstrated by decreases in the risetimes and fall times of unit hydrographs, and by multiplication of their flood peaks by a factor of two or more.

Along with the growth of impervious surfaces, urbanization causes a reduction in infiltration. This affects baseflow, which, since it is a longer-term part of the streamflow regime, does not produce the pronounced effects illustrated by surface runoff. Baseflow may increase in some urbanized streams due to residents watering lawns, sewage effluent or other deliberate transfers of water from other watersheds.

2.6 Exercises

Considering an urban area with which you are very familiar:

1 Do you think that the present storm hydrograph from this urban area is significantly different from the hydrograph from a similar storm from the pre-urbanized period? How would the storm hydrograph differ? Draw some hydrographs to demonstrate the change presented by the increased imperviousness.

2 Draw hydrographs for a hypothetical storm at selected points along a stream that drains the urban area. Explain the differences between the hydrographs.

3 Draw a model similar to Figure 2.3 describing how you think infiltration, depression storage, sub- and surface runoff, etc. have been affected by urbanization within the urban area.

References

ANDERSON, D. G. "Effect of Urban Development on Floods in Northern Virginia," U.S. Geological Survey Water Supply Paper No. 2001C, 22 pp. (1970).

BARRY, R. G. "The World Hydrological Cycle," in *Water, Earth and Man*. R. J. Chorley, ed. London:Methuen and Co., Ltd., pp. 11–29 (1969a).

BARRY, R. G. "Evaporation and Transpiration," in *Water, Earth and Man*. R. J. Chorley, ed. London:Methuen and Co., Ltd., pp. 169–184 (1969b).

BLACK, P. E. "Runoff from Watershed Models," *Water Resources Res.*, 6(2):465–477 (1970).

BRATER, E. F. and S. Sangal. "Effects of Urbanization on Peak Flows," in *Effects of Watershed Changes on Streamflow*. W. L. Moore and C. W. Morgan, eds. Austin:University of Texas Press, pp. 201–214 (1969).

References

CHOW, V. T. "Runoff," in *Handbook of Applied Hydrology*. V. T. Chow, ed. New York:McGraw-Hill Book Company, pp. 14-1–14-54 (1964).

CRIPPEN, J. R. "Changes in Character of Unit Hydrographs, Sharon Creek, California, after Suburban Development," U. S. Geological Survey Professional Paper No. 525-D, pp. 196–198 (1965).

DAVIS, S. N. and R. J. M. DeWiest. *Hydrogeology*. New York:John Wiley and Sons, Inc., 446 pp. (1966).

ESPEY, W. H. et al. "Urban Effects on the Unit Hydrograph," in *Effects of Watershed Changes on Streamflow*. W. L. Moore and C. W. Morgan, eds. Austin:University of Texas Press, pp. 215–228 (1969).

FLEMING, G. *Computer Simulation Techniques in Hydrology*. New York:Elsevier Publishing Co., 333 pp. (1975).

FLOHN, H. *Climate and Weather*. New York:McGraw Hill Book Co., 253 pp. (1969).

GILMAN, C. S. "Rainfall," in *Handbook of Applied Hydrology*. V. T. Chow, ed. New York: McGraw-Hill Book Co. (1964).

HARRIS, E. E. and S. E. Rantz. "Effect of Urban Growth on Streamflow Regimen of Permanente Creek, Santa Clara County, California," U.S. Geological Survey Water Supply Paper No. 1591-B (1964).

HEWLETT, J. D. and J. D. Helvey. "Effects of Forest Clearfelling on the Storm Hydrograph," *Water Resources Res.*, 6(3):768–782 (1970).

KIRKBY, M. F. "Infiltration, Throughflow and Overland Flow," in *Water, Earth and Man*. R. J. Chorley, ed. London:Methuen and Co., Ltd., pp. 215–227 (1969).

LEOPOLD, L. B. "Hydrology for Urban Land Planning—A Guidebook on the Hydrologic Effects of Urban Land Use," U.S. Geological Survey Circular No. 554, 18 pp. (1968).

LINSLEY, R. K. et al. *Applied Hydrology*. New York:McGraw-Hill Book Co., 689 pp. (1949).

NASH, J. E. "Applied Flood Hydrology," in *River Engineering and Water Conservation Works*. R. B. Thorn, ed. London:Butterworths, pp. 63–110 (1966).

PEIXOTO, J. P. and M. A. Kettani. "The Control of the Water Cycle," *Scientific Am.*, 228(4):46–61 (1973).

PHILIP, J. R. "The Theory of Infiltration: 4. Sorptivity and Algebraic Infiltration Equations," *Soil Sci.*, 84(3):257–264 (1957).

ROBERTS, M. C. and P. C. Klingeman. "The Influence of Landform and Precipitation Parameters on Flood Hydrographs," *J. Hydrology*, 11:393–411 (1970).

SAVINI, J. and J. C. Kammerer. "Urban Growth and the Water Regimen," U.S. Geological Survey Water Supply Paper No. 1591-A, pp. A-1–A-9 (1961).

SAWYER, R. M. "Effect of Urbanization on Storm Discharge and Ground Water Recharge in Nassau County, N.Y.," U.S. Geological Survey Professional Paper No. 475-C, Article 106, pp. C185–C187 (1963).

SHERMAN, L. K. "Streamflow from Rainfall by the Unit-Graph Method," *Eng. News-Rec.*, 108: 501–505 (1932).

SINGH, V. P. and S. Buapeng. "Effect of Rainfall-Excess Determination on Runoff Computation," *Water Resources Bull.*, 13(3):499–514 (1977).

STANKOWSKI, S. J. "Population Density as an Indirect Indicator of Urban and Suburban Land-Surface Modifications," U.S. Geological Survey Professional Paper No. 800-B, pp. B219–B224 (1972).

U.S. DEPARTMENT OF AGRICULTURE, FOREST SERVICE. "Municipal Watershed Management," in

Symposium Proceedings U.S.D.A., Northeastern Forest Experiment Station, Upper Darby, PA, 196 pp. (1975).

VIESSMAN, W., Jr. et al. *Introduction to Hydrology*. New York:Harper and Row Publishers, 704 pp. (1977).

Selected Readings

Aron, G. et al. "Infiltration Formula Based on SCS Curve Number," *J. Irr. Div., ASCE*, 103(IR4): 419–427 (1977).

Betson, R. P. "What Is Watershed Runoff?" *J. Geophys. Res.*, 69:1541–1552 (1964).

Boughton, W. C. "Hydrological Studies of Changes in Land Use," *Soil & Water*, 4:19–23 (1968).

Bras, R. L. and F. E. Perkins. "Effects of Urbanization on Catchment Response," *J. Hyd. Div., ASCE*, HY3:451–465 (1975).

Brater, E. F. "Steps Toward a Better Understanding of Urban Runoff Processes," *Water Resources Res.*, 4(2):335–347 (1968).

Brustkern, R. L. and H. J. Morel-Seytoux. "Analytical Treatment of Two-Phase Infiltration," *J. Hyd. Div., ASCE*, 96(HY12):2535–2548 (1970).

Cech, I. and K. Assof. "Quantitative Assessment of Changes in Urban Runoff," *J. Irr. Div., ASCE*, 102(IR1):119–126 (1976).

Crippen, J. R. "Selected Effects of Suburban Development on Runoff in a Small Basin Near Palo Alto, California," U.S. Geological Survey Open File Report, 19 pp. (1966).

Dean, J. D. and W. M. Snyder. "Temporally and Areally Distributed Rainfall," *J. Irr. Div., ASCE*, 103(IR2):221–229 (1977).

Department of State. "U.S. National Report on the Human Environment," Department of State Publication No. 8588, U.S. Government Printing Office (1971).

Dragoun, F. J. "Effects of Cultivation and Grass on Surface Runoff," *Water Resources Res.*, 5(5): 1078–1083 (1969).

Gray, D. M. "Interrelationships of Watershed Characteristics," *J. Geophys. Res.*, 66(4):1215–1223 (1961).

Hibbert, A. R. "Water Yield Changes after Converting a Forested Catchment to Grass," *Water Resources Res.*, 5(3):634–640 (1969).

Hills, R. C. "The Influence of Land Management and Soil Characteristics on Infiltration and the Occurrence of Overland Flow," *J. Hydrology*, 13:163–181 (1971).

Holtan, H. N. and D. E. Overton. "Analyses and Application of Simple Hydrographs," *J. Hydrology*, 1:250–264 (1963).

Holzman, B. "Sources of Moisture for Precipitation for the United States," U.S. Department of Agriculture, Technical Bulletin No. 589, U.S. Government Printing Office, 41 pp. (1937).

Hornbeck, J. W. et al. "Streamflow Changes after Forest Clearing in New England," *Water Resources Res.*, 6:1124–1131 (1970).

Jones, D. E., Jr. "Urban Hydrology: A Redirection," *Civil Eng.*, 37(8):58–62 (1967).

Jones, D. E., Jr. "Where Is Urban Hydrology Practice Today?" *J. Hyd. Div, ASCE*, 97(HY2): 257–264 (1971).

Komura, S. "Hydraulics of Slope Erosion by Overland Flow," *J. Hyd. Div., ASCE*, 102(HY10): 1573–1586 (1976).

Langbein, W. B. and K. T. Iseri. "General Introduction and Hydrologic Definitions," U.S. Geological Survey, Water Supply Paper No. 1541A (1960).
McPherson, M. B. "Hydrologic Effects of Urbanization in the U.S.," ASCE, Urban Water Resources Research Program, Technical Memo No. 17 (1972).
McPherson, M. B. "Urban Runoff," ASCE, Urban Water Resources Research Program, Technical Memo No. 18 (1972).
McPherson, M. B. *Hydrological Effects of Urbanization.* Paris:UNESCO Press, 280 pp. (1974).
Miller, C. R. and W. Viessman, Jr. "Runoff Volumes from Small Urban Watersheds," *Water Resources Res.*, 8(2):429–434 (1972).
Mitchell, W. D. "Effect of Artificial Storage on Peak Flow," U.S. Geological Survey Professional Paper 424-B, pp. 12–13 (1961).
Morel-Seytoux, J. H. and J. Khanji. "Derivation of an Equation of Infiltration," *Water Resources Res.*, 10(4):795–800 (1974).
Morel-Seytoux, J. H. and J. Khanji. "Equation of Infiltration with Comparison and Counterflow Effects," *Hyd. Sci. Bull.*, 20(3):505–517 (1975).
Nassif, S. H. and E. M. Wilson. "The Influence of Slope and Rain Intensity on Runoff and Infiltration," *Hyd. Sci. Bull.*, 20(4):539–553 (1975).
Petersen, M. S. *River Engineering.* Englewood Cliffs, NJ:Prentice-Hall Inc., 580 pp. (1986).
Philip, J. R. "Theory of Infiltration," in *Advances in Hydroscience, Vol. 5.* V. T. Chow, ed. New York:Academic Press, Inc., pp. 216–296 (1968).
Pluhowski, E. J. "Urbanization and Its Effect on the Temperature of the Streams on Long Island, N.Y.," U.S. Geological Survey Professional Paper 627-D, 110 pp. (1970).
Pluhowski, E. J. and A. G. Spinello. "Impact of Sewerage Systems on Stream Baseflow and Groundwater Recharge on Long Island, New York," *J. Res. U.S. Geological Survey*, 6(2):263–271 (1978).
Riggs, H. C. "Effect of Land Use on the Low Flow of Streams in Rappahannock County, Virginia," U.S. Geological Survey Professional Paper 525-C, pp. C196–C198 (1965).
Rothacher, J. "Increase in Water Yield Following Clear Cut Logging in the Pacific Northwest," *Water Resources Res.*, 6(2):653–658 (1970).
Schaake, J. C., Jr. "Water and the City," in *Urbanization and Environment.* T. R. Detwyler and M. G. Marcus, eds. Belmont, CA:Dubury Press, Chapter 5 (1972).
Schneider, W. J. and J. E. Goddard. "Extent and Development of Urban Flood Plains," U.S. Geological Survey, Circular No. 601-J, 14 pp. (1974).
Seaburn, G. E. "Effects of Urban Development on Direct Runoff to East Meadow Brook, Nassau County, New York," U.S. Geological Survey, Professional Paper 627-B, 14 pp. (1969).
Seaburn, G. E. "Preliminary Analysis of Rate of Movement of Storm Runoff Through Zone of Aeration Beneath a Recharge Basin on Long Island," U.S. Geological Survey Professional Paper 700-B, pp. 196–198 (1970).
Seaburn, G. E. and D. A. Aronson. "Influence of Recharge Basins on the Hydrology of Nassau and Suffolk Counties, Long Island, N.Y.," U.S. Geological Survey, Water Supply Paper 2031, 66 pp. (1974).
Shen, H. W. and R. M. Li. "Rainfall Effect on Sheet Flow Over Smooth Surface," *J. Hyd. Div., ASCE*, 99(HY5):771–792 (1973).
Singh, V. P. "A Laboratory Investigation of Surface Runoff," *J. Hydrology*, 25:187–200 (1975).
Smith, R. E. and D. A. Woolhiser. "Overland Flow on an Infiltrating Surface," *Water Resources Res.*, 7(4):899–913 (1971).

Soil Conservation Service. "Hydrology," Section 4 in *SCS National Engineering Handbook*. U.S. Department of Agriculture, U.S. Government Printing Office (1971).

Soil Conservation Service. "Urban Hydrology for Small Watersheds," Technical Release No. 55 (1975).

Task Force on the Effect of Urban Development on Flood Discharges, Committee on Flood Control. "Effect of Urban Development on Flood Discharges: Current Knowledge and Future Needs," *J. Hyd. Div., ASCE*, 95:287–309 (1969).

Tholin, A. L. and C. J. Keifer. "The Hydrology of Urban Runoff," *Trans. ASC Eng.*, 125:1308–1373 (1960).

Thomas, H. E. and W. J. Schneider. "Water as an Urban Resource and Nuisance," U.S. Geological Survey Circular 601-D (1970).

Viessman, W., Jr. "The Hydrology of Small Impervious Areas," *Water Resources Res.*, 2(3):405–412 (1966).

Viessman, W., Jr. et al. "Urban Storm Runoff Relations," *Water Resources Res.*, 6(1):275–279 (1970).

Waananen, A. O. "Hydrologic Effects of Urban Growth: Some Characteristics of Urban Runoff," U.S. Geological Survey Professional Paper 424-C, pp. C353–C356 (1961).

Waananen, A. O. "Urban Effects on Water Yield," in *Effects of Watershed Changes on Streamflow*. Walter L. Moore and C. W. Morgan, eds. Austin:University of Texas Press, pp. 169–182 (1969).

Wallace, J. R. "The Effects of Land Use on the Hydrology of an Urban Watershed," Report No. ERC-0871, School of Civil Engineering, Georgia Institute of Technology, Atlanta, 66 pp. (1971).

Weidner, C. H. *Water for a City*. New Brunswich, NJ:Rutgers University Press, 339 pp. (1974).

Woolhiser, D. A. "Overland Flow on a Converging Surface," *Trans. ASAE*, 12(4):460–462 (1969).

Woolhiser, D. A. and J. A. Liggett. "Unsteady, One-Dimensional Flow Over a Plane: The Rising Hydrograph," *Water Resources Res.*, 3(3):753–771 (1967).

3 Urbanization and Stream Water Quality

. . . simply stated, the problem is as follows. When a city takes a bath, what do you do with the dirty water? – Field and Lager (1975), p. 107.

3.1 Water Quality of Natural Streams

THE MESHING OF WATER QUANTITY AND WATER QUALITY

THE VARIOUS HYDROLOGIC and hydraulic variables which affect water quantity in a river system also influence system water quality. For example, velocity, depth and cross-sectional area along a stream are directly related to the underlying geology. In the process of erosion of the underlying geology, many (geo)chemical constituents are added to the river system. Runoff also contributes inorganic and organic constituents. Changes in water quantity such as flood peaks or droughts cause dilutions of the constituents to vary and alter water quality. In a geologic time sense, these changes in water quality occur in a moment; however, they can seriously modify the biota along a river system, because only a brief time is required to bring about massive kills of flora and fauna.

All these short-term water quantity/water quality interactions are ultimately controlled by the much longer termed regional climatic and geologic history (Figure 3.1). Interactions do occur between the geology and climate at a particular time and between the water quantity and water quality. For example, certain dissolved constituents carried by the water and given impetus by the climate (precipitation) could cause the water to be more erosive than at another moment in geologic time. These waters could temporarily bring about drastic water quality changes and also cause enormous downcutting of the stream channels.

Therefore, when a researcher studies a particular river system, he must be aware of the meshing of water quantity and water quality with geology and climate. These factors all interact to produce a river which is unique, and which can only be grossly compared to other rivers. Each river makes its own imprint on the geology and in general has a personality of its own.[12]

The longitudinal profile of a stream is an imprint which developed in response to the preceding factors. Each profile is unique. However, in general, longitudinal profiles of many rivers have been shown to be concave to the sky (see Figure 3.2; Hack, 1957; Morisawa, 1968). This profile represents an adjustment of the various hydraulic variables for the efficient utilization of energy within that river system (Langbein and Leopold, 1964). A typical longitudinal description for a large river is a stream which has its headwaters in the mountains, and from there it flows swiftly over rocks and boulders to the foothills, where the slopes become smoother and the velocities decrease (Figure 3.2). In the coastal plains, the stream begins meandering and velocities diminish further, allowing the deposition of sand and silt to occur.

Velz (1970) suggests that for the purpose of stream sanitation there are two broad classifications of streams: (1) the pool and riffle, which has been partially explained above; and (2) the regular gradient type, which occurs in the lower reaches of many streams. Further elaborating on Velz and observing the longitudinal profile more closely, one will find nicks (Figure 3.2) most likely caused by geologic discontinuities, i.e., changes in geology. These nicks are indicative of riffles (shallow reaches) and pools (deep reaches), and are of prime importance to water quality. At riffles, mechanical aeration is enhanced whereby dissolved oxygen supply is increased. In the pools, velocities are much slower than in the riffles, and a different type of biota may develop.

Riffles and pools together are integral parts of the ecological makeup of a particular river system. Hynes (1960) points out that the speed of the current and the amount of dissolved oxygen in a stream are two critical factors in the development of a habitat for many different species of aquatic flora and fauna.

The preceding has been a very generalized description of natural river systems. From this description, one can develop a generalized definition of natural stream water quality. Hem (1970) observed:

> . . . the chemical composition of natural water is derived from many different sources of solutes, including gases and aerosols from the atmosphere, weathering and erosion of rocks and soils, solution or precipitation reactions occurring below the land surface, and cultural effects resulting from activities of man (p. 1).

We can see by the last line of this statement that so-called "natural" water has been affected by man. In reality, many natural streams have been affected by man through damming, channelization, addition of various inorganic and/or organic chemicals and other activities.

---

[12]In trying to arrive at a description of a natural stream, an aquatic biologist, Hynes (1960), came to the conclusion that ". . . rivers are strict individualists, each of which varies in its own way so as to make nonsense of anything but a very broad general classification" (p. 15).

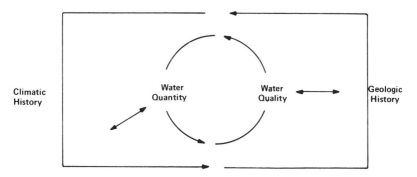

Figure 3.1. The control of water quantity and quality by climatic and geologic history.

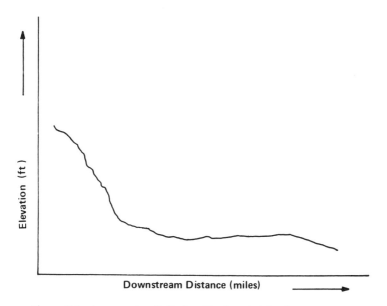

Figure 3.2. A concave longitudinal profile characteristic of many streams.

In defining "natural" or "free" water, one must realize that it is not (any longer) "pure" water, at least not by ancient standards. Free water is, in fact, a dilute aqueous solution of a mixed and complex character.

WATER QUALITY OF NATURAL RIVERS

Table 3.1 lists selected chemical concentrations found in the Amazon River, the Mississippi River and, in the third column, the mean composition of world river water. It can be seen from this table that river water is indeed a mixture containing many solutes. This mixture has developed over time and is directly related to the geochemical, vegetative and climatic environment through which the stream passes.

Table 3.1. Selected chemical concentrations in river water.

Constituent	1[a] July 16, 1963		2[b] Oct. 1, 1962– Sept. 30, 1963		3[c]	
	mg/l	meq/l	mg/l	meq/l	mg/l	meq/l
Silica (SiO_2)	7.0		6.7		13	
Aluminum (Al)	0.07					
Iron (Fe)	0.06		0.04		0.67	
Calcium (Ca)	4.3	0.214	42	2.096	15	0.750
Magnesium (Mg)	1.1	0.090	12	0.987	4.1	0.342
Sodium (Na)	1.8	0.078	25	1.088	6.3	0.274
Potassium (K)			2.9	0.074	2.3	0.059
Bicarbonate (HCO_3)	19	0.311	132	2.163	58	0.958
Sulfate (SO_4)	3.0	0.062	56	1.166	11	0.233
Chloride (Cl)	1.9	0.054	30	0.846	7.8	0.220
Fluoride (F)	0.2	0.011	0.2	0.011		
Nitrate (NO_3)	0.1	0.002	2.4	0.039	1	0.017
Dissolved Solids	28		256		90	
Hardness as $CaCO_3$	15		155		55	
Noncarbonate	0		47		7	
Specific Conductance (micromhos at 25°C)	40		421			
pH	6.5		7.5			
Color			10			
Dissolved Oxygen	5.8					
Temperature °C	28.4					

[a]Amazon at Obidos, Brazil. Discharge 7,640,000 cfs (high stage).
[b]Mississippi at Luling Ferry, LA (17 mi west of New Orleans). Time-weighted mean of daily samples.
[c]Mean composition of river water of the world.
Date under sample number is date of collection. Source of data: 1, Oltman (1968), p. 13; 2, U.S. Geol. Survey Water-Supply Paper (1950) p. 316; 3, Livingstone (1963), p. 41.
Source: Hem (1970), p. 12.

Biesecker and Leifeste (1975) analyzed water quality data from 57 hydrologic benchmark stations located in 37 states. Each station was carefully selected to be closely representative of a "natural" stream for that respective region. They developed median curves showing the relationship between dissolved solids concentration and water discharge per unit area for any natural stream in 11 of the 14 physical divisions of the U.S. The maximum dissolved solids concentration for selected streams is shown in Table 3.2. Nitrate concentration is shown by water resources region in Table 3.3. Pesticide data collected from the benchmark stations revealed that the DDT family accounted for 75% of the occurrences. The highest occurrence was about 100 times below the permissible maximum recommended by the U.S. Federal Water Pollution Control Administration (1968). Toxic minor metals concentrations were also very low. About 65% of the 645 measurements showed near zero concentrations. Three measurements indicated concentrations in excess of drinking water standards recommended by the U.S. Public Health Service (1962).

HABITAT DISTRIBUTION IN A NATURAL STREAM[13]

Huet (1954) divided a natural river into four biotic zones:

(1) The trout zone—rapidly flowing, cool, well-oxygenated water and steep-sided canyons and valleys
(2) The grayling zone—less rapid velocities which may be slow in some reaches, where the water is always well-oxygenated as it flows through steep-sided to flat-bottomed valleys
(3) The barbel zone—moderate current with long slow reaches, where dissolved oxygen may decrease in concentration in warm weather, with wide, flat-bottomed valleys
(4) The bream zone—where water flows sluggishly, the streams are canal-like, oxygen content may fall to low levels and only hardy fish survive

It can be seen from the preceding that stream fauna carefully adjust to the water mixture character, dissolved oxygen concentrations and flow velocities. Eaton (1977) points out three features of organism communities which may be used to determine whether a river is unpolluted. These are diversity,[14] stability and self-purification. A wide diversity of plant and animal species is indicative of a healthy river community. A few minutes of collecting should yield up to two dozen visible species, and under a microscope the total number of species of microorganisms might approach 100. This is largely due to the favorable environment that most natural rivers provide. The water current furnishes a con-

---

[13]See Usinger (1967) for illustrations and discussions of habitat distribution.
[14]See Armstrong et al. (1968) for an excellent discussion on the species diversity index.

Table 3.2. Selected hydrologic characteristics, hydrologic benchmark stations.

Station No.	State	Drainage Area (mi²)	Principal Rock Type	Average Annual Runoff (in.)	Maximum Dissolved-Solids Concentration[a] (mg/l)
1	Alabama	86.8		23	N.D.[b]
2	Alabama	91.3		26	70
3	Arizona	36.4	Granite	2	258
4	Arkansas	89.4	Shale, Sandstone	23	48
5	Arkansas	58.4	Limestone	19	186
6	California	6.5	Marine Sedimentary	50	88
7	California	181	Granite	25	27
8	California	23.7	Metamorphized Sedimentary Rocks	0.02	N.D.
9	Colorado	23	Schist	18	61
10	Colorado	72.1	Schist, Conglomerate	20	51
11	Florida	97.9	Limestone	25	153
12	Georgia	72.2	Gneiss, Schist	14	103
13	Georgia	56.5	Gneiss, Schist	40	20
14	Hawaii	11.6	Volcanic	N.D.	N.D.
15	Idaho	22.0	Quartzite	21	60
16	Idaho	253	Volcanic Rocks	0.1	140
17	Indiana	38.2	Limestone	12	353
18	Iowa	52.5	Limestone, Shale	6	506
19	Louisiana	51	Unconsolidated Sand	16	48
20	Maine	69.5	Gneiss, Schist	34	23
21	Michigan	13.6	Sandstone	13	130
22	Minnesota	253	Granite, Gabbro	10	31
23	Minnesota	101	Limestone, Sandstone	4	314
24	Mississippi	52.2	Unconsolidated Sand	20	29
25	Montana	100	Shale, Sandstone	2–3	1,870
26	Montana	31.4	Limestone, Quartzite	64	56
27	Nebraska	960	Unconsolidated Sand	3	172
28	Nevada	20	Volcanic	3	94
29	Nevada	11.1	Limestone	5	203

(continued)

Table 3.2. (continued).

Station No.	State	Drainage Area (mi²)	Principal Rock Type	Average Annual Runoff (in.)	Maximum Dissolved-Solids Concentration[a] (mg/l)
30	New Jersey	2.31	Unconsolidated Sand	13	25
31	New Mexico	69	Volcanic Rocks	3.5	110
32	New Mexico	53.2		7	72
33	New York	59.5	Sandstone	25	44
34	North Carolina	49.2	Sandstone	30	20
35	North Dakota	74	Sandstone, Silt	1	3,420
36	North Dakota	160	Shale, Till	<1	1,220
37	Ohio	12.8	Shale, Sandstone	15	83
38	Oklahoma	24.6	Granite	3–4	213
39	Oklahoma	40.1	Shale, Sandstone	20–25	29
40	Oregon	26.2	Andesite	Lake	
41	Oregon	240	Volcanic Rocks	25	57
42	Pennsylvania	46.2	Shale, Sandstone	17	36
43	South Carolina	70	Unconsolidated Sand	15–20	29
44	South Carolina	87	Unconsolidated Sand	14	17
45	South Dakota	51	Shale, Schist	1.5	292
46	South Dakota	51	Till	1	N.D.
47	Tennessee	447	Chert	22	65
48	Tennessee	106	Shale, Sandstone	36	19
49	Texas	52.4	Volcanic Rocks	0.15	126
50	Texas	34.2	Limestone, Marl	3	291
51	Utah	7.25		9–10	447
52	Virginia	8.53	Schist, Quartzite	13–14	42
53	Washington	22.1	Granite	N.D.	N.D.
54	Washington	74.1	Slate	145	60
55	Wisconsin	131	Schist, Granite	12	152
56	Wyoming	10	Shale, Sandstone	19	207
57	Wyoming	72.7	Granite	16–18	56

[a]Maximum observed during 1968–70 water years; monthly sampling.
[b]N.D. = not determined.
Source: Biesecker and Leifeste (1975), pp. 17–18.

Table 3.3. Nitrate concentration of hydrologic benchmark stations and selected major streams draining various water-resources regions, 1968–70 water years.

Water-Resources Region	Nitrate Concentration (mg/l)			
	Hydrologic Benchmarks		Selected Major Streams	
	Average Maximum	Average Median	Average Maximum	Average Median
North Atlantic	1.7	0.3	5.9	2.9
South Atlantic—Gulf	1.7	0.5	2.1	0.4
Missouri	4.9	0.4	7.4	1.0
Arkansas—White-Red	1.0	0.2	16	1.9
Columbia—North Pacific	1.0	0.2	6.1	1.0

Source: Biesecker & Leifeste (1975), p. 15.

stant stream of organic and inorganic particles and nutrients for the plants and animals, as well as a continuous supply of life-maintaining water and dissolved oxygen.

The wide diversity of flora and fauna encourages a stability in river communities. An ecological food web is formed which is complex and self-regulatory. If one species loses numbers for some reason, the number of its predators will also be reduced, and the species will be allowed to grow and once again regain its original number. This self-balancing is demonstrated by the amazing ability of a river community to rapidly reestablish itself to original population levels after a major flood.

The third factor described by Eaton is self-purification. He points out that rivers do indeed receive large amounts of organic debris from the lands they drain. In addition, the aquatic community contributes a load to the river when individuals die and decay. This material enters the food web, and is rapidly and completely degraded.

Langbein and Durum (1967) investigated the aeration capacity of streams. They observed:

> ... the oxygenation of a stream is a function of the biologic, physical and hydraulic properties of the stream. Oxygen may be added by such processes as photosynthesis of aquatic vegetation, and by mechanical aeration of the flowing water.

They expressed the rate of absorption of oxygen per unit of time by a simple equation:

$$dc/dt = k_2(c_s - c)$$

where

c = the concentration of oxygen (mg/l)
c_s = the concentration for saturation at a given temperature
t = the time in days
k_2 = the coefficient of reaeration [derived from the second coefficient in the Streeter-Phelps (1925) formulation]

Langbein and Durum go on to point out that the hydraulic properties of a given river have a significant effect on the reaeration coefficient (k_2). This relationship may be approximated by the equation:

$$k_2 = 3.3 v/H^{1.33}$$

where

v = mean velocity of the stream (ft/sec)
H = the mean depth (ft)

When this formula is applied to data from several regions, it is found that coastal plain streams with smooth slopes, located in populous areas have reaeration capacities which are low relative to their discharge rates.

It must be noted that parts of natural rivers may be "naturally" polluted.[15] Hynes (1960) pointed out that ". . . in densely wooded regions, the autumn leaf-fall may add so much organic matter to water that fish are asphyxiated" (p. 1). Naturally flowing sluggish woodland streams are excellent examples of naturally polluted water. Certain inorganic chemicals coming from the geochemistry of a particular region may appear in the river waters and effectively pollute them. For example, Hem (1970) pointed out that in New Zealand, deaths among cattle have resulted from drinking natural water containing arsenic.

3.2 Impact of Raw Sewage on Stream Water Quality

Bartsch (1948) wrote:

> . . . the entry of pollutants into a flowing stream sets off a progressive series of physical, chemical and biological events in the downstream waters. Their nature is governed by the character and quantity of the polluting substance (p. 13).

---

[15]Pollution is defined after Stoker and Seager (1972) as a departure from a normal rather than a pure state. Normal water is defined as having no substances in sufficient concentrations as to prevent the water from being used for purposes thought of as normal.

To illustrate the gross effects of pollutants on a stream system, in this section we will briefly discuss the impacts of raw sewage discharge into a hypothetical river.

After entry into the stream waters, sewage[16] acts as an excellent food source for bacteria, and logarithmically stimulates their growth. As they multiply, they require large amounts of dissolved oxygen, i.e., they exert a high biochemical oxygen demand (BOD) and seriously draw upon the stream's supply of dissolved oxygen.

Stoker and Seager (1972) demonstrate the impact of a heavy BOD loading created by carbon-bearing pollution (of which sewage is a prime example) by the simple reaction:

$$C + O_2 \rightarrow CO_2$$

With bacterial help, the oxidation of C to CO_2 occurs, and the reaction proceeds from left to right (see Figure 3.3). In this process, 32 g of oxygen are required to oxidize 12 g of carbon. As Stoker and Seager illustrate, ". . . this amounts to a reaction between the dissolved oxygen from a gallon (3.8 liters) of water and a small drop of oil" (p. 92). Hence, dissolved oxygen may rapidly be depleted in this manner.

If one were to take dissolved oxygen measurements at various points along a river progressing downstream, one would derive a dissolved oxygen sag curve as shown in Figure 3.4. From before the point of sewage entry (at mile 0) to mile 15 downstream, bacterial decomposition proceeds mainly by aerobiosis. From mile 15 to about mile 30, bacterial decomposition occurs by means of anaerobiosis; after this point, aerobic conditions return. We will now briefly discuss the chemistry involved.

AEROBIC BACTERIAL DECOMPOSITION

Turk et al. (1974) defined aerobiosis as bacterial decomposition in the presence of air. For a given weight of nutrients, this process yields the most energy as shown by the following reaction.

$$C_xH_yO_zN_qS + O_2 \rightarrow CO_2 + H_2O + NH_4^+ + SO_4^{-2} + 4100 \text{ cal/g of protein}$$

generalized protein molecule — ammonium ion — sulfate ion

The reaction above is typical of the first stage of bacterial deoxygenation of polluted waters. This is called the "carbonaceous" stage. The BOD exerted

---

[16]See Painter (1971), pp. 334–348, for a description of raw sewage.

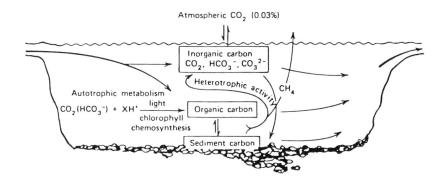

Figure 3.3. Schematic representation of the carbon cycle (*Source*: Kerr in: *Nutrients in Natural Waters*, Allen and Kramer, eds. © 1972 by Wiley-Interscience et al. reprinted by permission of John Wiley and Sons, Inc.).

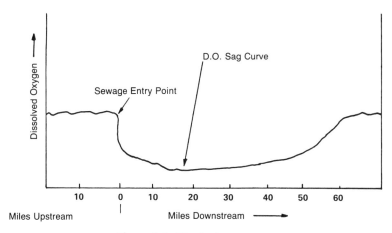

Figure 3.4. Dissolved oxygen sag curve.

upon the receiving body during this stage is calculated by this formula (Zanoni, 1967):

$$Y_t = L(1 - 10^{-kt})$$

where

Y_t = oxygen used in (t) days expressed in mg/l (usually referred to as BOD)
k = first-order velocity constant with units of day^{-1} (water temperature is determining factor)
L = ultimate carbonaceous or first-stage demand (mg/l)

After this stage is accomplished and the organic nutrient exhausted, other aerobic bacteria facilitate the oxidation of ammonium salts to derive additional energy as depicted in the following example.

$$NH_4^+ + 2O_2 \rightarrow 2H^+ + H_2O + NO_3^- + 4350 \text{ cal/g of ammonium ion}$$

ammonium ion   nitrate ion

This process is called nitrification (Turk et al., 1974) or the second stage of BOD exertion, and proceeds at a rate which is independent of the first stage.

Courchaine (1968) studied nitrogenous BOD (nitrification) in the Grand River at Lansing, Michigan (Figure 3.5). His results demonstrated, in contrast to those of Zanoni and others, that carbonaceous and nitrogenous BOD occurred *simultaneously*; however, he found that the rates differed. For the first six days, the oxidation rate of carbonaceous matter exceeded nitrification, and thereafter the nitrification rate predominated (Figure 3.6). Courchaine pointed out that the nitrogenous BOD curve could be divided into two time periods or phases: (1) the lag phase, in which the bacterial cells undergo multiplications, building up to a maximum population and (2) the period after the initial buildup.

The explanation proposed for this lag phase has to do with the energy requirements of the specific bacteria involved. Heterotrophic bacteria obtain their energy from organic matter and carry out the oxidation of the carbonaceous material. As Courchaine points out, they consist of a wide variety of individual types of bacteria with optimum living temperatures ranging from 18°C to 25°C.

As shown in Figure 3.5, nitrogenous oxidation is carried out by aerobic bacteria which obtain their energy from the oxidation of ammonia and its derivatives. These are called autotrophic bacteria, and utilize inorganic compounds as in the following formulae (after Ruane and Krenkel, 1978):

$$NH_4^+ + (3/2)O_2 \xrightarrow{\text{nitrosomonas}} NO_2^- + 2H^+ + H_2O$$

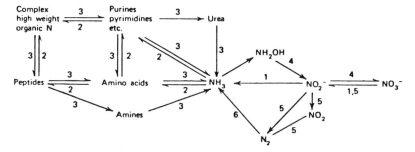

Figure 3.5. Simplified nitrogen cycle showing main molecular transformations: 1. nitrate assimilation; 2. ammonia assimilation; 3. ammonification; 4. nitrification; 5. denitrification; 6. nitrogen fixation (*Source*: Brezonik in: *Nutrients in Natural Waters*. Allen and Kramer, eds. © 1972 by Wiley-Interscience et al., reprinted by permission of John Wiley and Sons, Inc., p. 2).

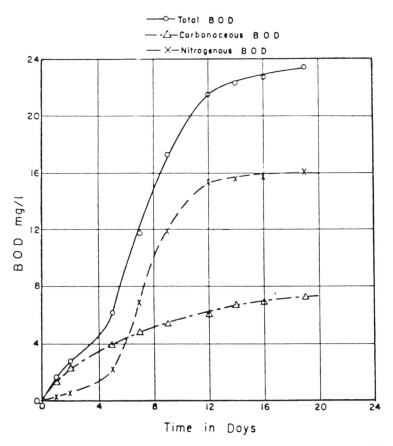

Figure 3.6. Grand River BOD curve, Canal Road–Delta Mills [*Source*: Courchaine (1968) p. 837].

and

$$NO_2^- + (1/2)O_2 \xrightarrow{\text{nitrobacter}} NO_3^-$$

Nitrosomonas and nitrobacter have optimum living temperatures for growth of 25°C to 28°C. Courchaine observed that temperatures at his sampling points averaged between 25°C and 28°C,[17] the optimum range for nitrifying bacteria. Hence, he attributes this as being a prime cause for the lag phase and the simultaneous BOD created by carbonaceous and nitrifying bacterial oxidation. Another factor of possibly equal importance is the presence of a viable population of nitrifiers. Courchaine mentions that some streams might lack the latter, and that this might partially explain why researchers in the past have observed two separate BOD exertion phases.

ANAEROBIC BACTERIAL DECOMPOSITION

In the region between mile 15 and mile 40 on Figure 3.4, dissolved oxygen is depleted and bacterial decomposition occurs by means of anaerobiosis. As Turk et al. (1974) point out, this type of decomposition does not yield as much energy; however, it is still energetically profitable. Anaerobic decomposition of sugars and carbohydrates is called fermentation, and is demonstrated by the following reaction:

$$C_6H_{12}O_6 \rightarrow 2C_2H_6O + 2CO_2 + 100 \text{ gal/g of sugar}$$
$$\text{glucose} \qquad \text{ethyl alcohol}$$

Putrefaction is the anaerobic decomposition of proteins and may be demonstrated by the following reaction:

$$C_xH_yO_zN_qS + H_2O \rightarrow NH_4^+ + CO_2 + CH_4 + H_2S + 370 \text{ cal/g}$$
$$\text{of protein}$$
$$\text{a protein} \qquad\qquad\qquad \text{methane}$$

Carbohydrate matter may be converted to methane:[18]

$$C_6H_{12}O_6 \rightarrow 3CH_4 + 3CO_2 + 220 \text{ cal/g of sugar}$$

---

[17]Courchaine (1968) writes that the city of Lansing, through thermal discharges from the condensor cooling water from the steam power generating plants, added an average of 15 billion Btu/day during the August 1960 survey.

[18]Benoit (1971) observed that a broad class of anaerobic microbial reactions consists of the partitioning of organic matter into carbon dioxide and molecular species more reduced than carbohydrate, as shown in the following equation: $2CH_2O \rightarrow CO_2 + CH_4$ (methane).

Putrefaction makes water bubble and gives off foul odors, mainly from hydrogen sulfide. Methane gas accumulation in sewers may be hazardous since it is explosive. Needless to say, the aesthetics of such a reach of stream leaves much to be desired.

CHANGING BIOTIC CHARACTER

Water pollution chemistry has been briefly presented; now let us look at the easily observable and objectionable results of stream pollution. The narrative which follows is based on papers by Bartsch (1948) and Bartsch and Ingram (1950). From mile 0 to mile 40 on Figure 3.4, bacterial growth has been greatly enhanced by the introduction of sewage nutrients. The aquatic ecology of this reach is upset, and soon many of the former organisms find the new living conditions unsuitable and die or move to new habitats. Bacterial life predominates, and a region low in dissolved oxygen or a zone of degradation and active decomposition is created. At 10 miles below the city, dissolved oxygen is absent for the next 20 miles to mile 30. At these points, the BOD[19] is greater than the supply of oxygen provided by solution from the atmosphere (since photosynthesis from any remaining green plant life is almost nil). This is a septic region occupied by bacteria, bacteria-eating protozoans and other organisms which must find their oxygen by anaerobic reduction of oxygen-bearing compounds or by special adaptation. This is usually accomplished (as has been partially presented) by reducing nitrates and nitrites and sulfates.

Rattailed maggots, sludge worms, blood worms and several others are especially adapted for life in sludge. For example, the rattailed maggot has a snorkel-like telescopic air tube which it extends above the water surface to breathe atmospheric oxygen, thereby eliminating the necessity for dissolved oxygen. These animals are all pollution-tolerant, and by the same reasoning are indicators of polluted waters.

A predatory relationship exists between ciliated protozoa and bacteria. This is an important factor, because the bacteria work best when they are growing rapidly and do not achieve a stable lazy population level. The aggressive and ferocious protozoa maintain the bacterial population at levels where the bacteria are most effective at degrading sewage. Bartsch and Ingram claim this is one of the reasons why sewage disappears so effectively from streams.

Within this septic region, massive colonies of microorganisms usually accumulate. These are called sewage fungus by Hynes (1960). In reality, only a few of the organisms involved are fungi; some are bacteria and other animals. Colonial masses formed may be colored brown, white, yellow or pink, and may form a carpet over mud surfaces. When they die, they usually cause foul odors.

---

[19]Velz (1970) in Chapter 4 on Organic Self-Purification, discusses BOD as liabilities and assets, and in so doing presents an excellent description of the concept of BOD.

Sludge deposits on the stream bottom are greatest nearest the sewage discharge and decrease in thickness downstream. Sewage molds and various bacteria degrade this material. Invertebrates such as the rattailed maggot find the sludge a suitable habitat.

Few algae can occupy the degradation region due to the floating debris, suspended solids and consequent turbidity. Even algae of the slimy blue-green variety and bottom types are sparse in this reach.

As one progresses downstream and the dissolved oxygen sag curve begins recovery, the pollution indicators start to diminish in number. The ciliated protozoans overcome the bacteria and they in turn fall prey to the rotifers who fall prey to the crustaceans, and so on. Eventually, the pollution-tolerant plants and animals are suppressed, and the association which represented clean waters upstream returns. The water clears, suspensions of green algae and other plants reappear, and the processes of respiration, photosynthesis and reaeration provide for a healthy aquatic ecology.

3.3 Sources of Urban Pollutants

In general, sources of pollutants may be classified into two categories: point source or nonpoint source. An example of a point source would be the outlet into a river of a municipal sewerage system or an industry directly discharging effluent into a river. The "point" of origin of these pollutants may easily be observed and identified. A nonpoint source consists of pollutants derived from a broader area, and their origins are not at all easy to determine or, for that matter, to control. Examples are sediment or acid mine drainage or urban runoff (Whipple et al., 1978).

The federal Water Pollution Control Act as amended in 1972 (Public Law 92-500) has provided the funding and initiated the water quality management planning required to reduce urban and rural point source pollution to prescribed standards. This law and its impact will be discussed in Chapter 6, and will not be presented here. Point source pollution[20] also will not be discussed. Rather, the much more uncontrollable and, according to Whipple et al. (1974) and Radziul et al. (1975), greatest threat to urban water quality, nonpoint source pollution, will be presented.

In the past, the relatively sparse population inhabiting the earth was widely scattered. People could dispose of their organic and inorganic litter in almost any direction, and the natural elements could easily degrade and assimilate it. It has been shown that in the last two centuries there has occurred a tremendous movement and concentration of people into urban areas. At the same time, a similar growth of industrialization has occurred. These huge agglom-

---

[20]Point sources are discussed in many fine texts, e.g., Velz (1970) and Best and Ross (1977).

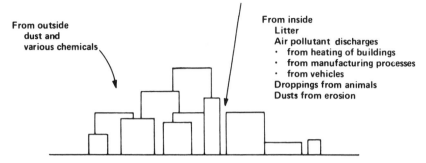

Figure 3.7. Origins of urban nonpoint source pollution.

erations have created the generation of mountains of litter which for proper degradation far exceed the abilities of the local environment. Additionally, industries and the industrial revolution have created new packaging materials which require a long time for degradation. All in all, the concentration and potency of urban litter have created environmentally and ecologically upsetting problems.

The American Public Works Association (1969) defined street refuse or litter as ". . . the accumulation of materials found on the street, sidewalk, or along the curb and gutter which can be removed by sweeping" (p. 35).[21] Litter includes remnants resulting from careless public and private waste collection operations, animal and bird droppings, soil washed or eroded from land surfaces, construction debris, road surfacing materials ravelled by travel, impact, frost action or other causes, air pollution dust falls, windblown dirt from open areas and a host of subsidiary materials.

Figure 3.7 illustrates the origins of urban nonpoint source pollution. From without, exotic dust enters the urban system. Some dust has been shown to have originated thousands of miles away from the point of deposition. Sartor and Boyd (1977) suggest:

> . . . a large fraction of the particulate matter contributing to the water pollution effects of street surface contaminants are of a size fine enough that they could have been transported by air currents prior to being deposited on the street surface (p. 9).

However, most likely the majority of urban dust is generated locally. Along with the dustfall may come various chemicals. Kellog et al. (1972) discuss the washout from the atmosphere of the various sulfur compounds.

Gambell and Fisher (1966) observed the composition of rainfall over a 34,000-mi² area in Southeastern Virginia and North Carolina. They studied

---

[21]For the purposes of this chapter, we will expand this definition to include refuse materials found not only on streets, but on all urban surfaces.

data on Na^+, Ca^{+2}, Mg^{+2}, NO_3^-, SO_4^{-2} and Cl^-. They observed that excluding bicarbonate, almost one-half of the dissolved solids carried by streams were contributed by rainfall; rainfall added more nitrates and sulfates to the streams than were carried by the streams. Durum (1971) suggested that precipitation contributed about one-quarter of the total chloride stream load. Therefore, substantial quantities of chemicals may be contributed by airborne exotics.

From within the urban areas come refuse, hydrocarbons and gases from heating, various particulates from manufacturing processes, carbon, sulfur and lead compounds from vehicles, animal droppings, leaves and dust from the erosion of urban and natural surfaces.

Weibel et al. (1966) and Loehr (1974) presented data concerning the constituents in urban rainfall. Weibel et al. found that nitrogen, phosphates, suspended solids and COD (chemical oxygen demand) were relatively low in comparison to concentrations found in sanitary sewage. Loehr pointed out that sulfate concentrations in precipitation were found to be greatest near industries in comparison to other urban locations. Laxen and Harrison (1977) discussed the airborne dispersal of lead contributed to highways and streets by automobiles. All in all, the chemical constituents of urban rainfall and air represent a significant quantity of pollutants which may mix with the larger parts of organic and inorganic matter constituting urban litter.[22]

Overton and Meadows (1976) divided urban land use patterns into five principal settings: (1) construction, (2) industry, (3) commerce, (4) streets and roads, and (5) residential sections. Construction activities have been shown to yield a brief but clearly demonstrable impact on stream water quality. Construction activities have several phases: (1) site preparation, i.e., clearing and grubbing, (2) dynamiting, (3) demolition, (4) excavation, (5) cut and fill operations and (6) compacting and handling materials. In site preparation, erosion may be greatly accelerated, and streams clogged with sediment (Guy, 1963; Wolman, 1968). In later phases, dust may be generated, oils and lubricants spilled from machinery, inorganics leaked or corroded from the building materials and organics released from litter and food wastes.

Many industries have a significant effect on storm water quality. Open stockpiles of raw materials, when exposed to rainfall, may release pollutants by leaking. Inadvertent spills from the handling of materials, leaking pipe systems, and storage units add to the problem. Process gases may contribute to the impairment of water quality through dustfall or washout. Industrial pollutants that could have a serious effect on the receiving water source include toxics, oils and heavy metals.

Commercial areas are usually highly impervious, consisting of buildings, streets and parking lots. Exposed surfaces may accumulate pollutants generated by automobile traffic, litter, dustfall and spills.

[22]Recent research has suggested that phosphorous and possibly other chemicals may adsorb on particles of sediment and be washed into streams (see Novotny et al. 1978).

Streets and roads accumulate dustfall, spills, littering and wastes from automobile use. Automobile use generates a number of pollutants: exhaust, emissions, tire wear and leakage of oils. From these sources may come lead, rubber compounds, oils and high COD.

Litter of all types and sizes may originate from residential areas. These pollutants cover the realm from the large-sized abandoned automobile to the microscopic organic coliform.

3.4 Impacts of Urban Runoff on Water Quality

THE MESHING OF URBAN WATER QUANTITY AND WATER QUALITY

Chapter 2 presented the increases in storm runoff created by the growth of impervious surfaces within urban areas. In Section 3.3, urban litter was discussed. This section will briefly discuss the relationship between the increases in peak flows and the expected stream water quality impairment.

Sartor and Boyd (1977) explained the separate mechanisms involved in washing surface contaminants (urban litter) off streets and their flow into receiving waters. They outline the process as: (1) freeing the contaminant from the street surface, (2) transporting the particles transversely across the surface to a gutter by overland, sheet-like flow, (3) carrying the particles parallel to the curb line to the storm sewer, and (4) carrying the particles through the storm sewer. They point out further that the contaminants are removed via two mechanisms which seem to operate simultaneously, (1) the soluble fractions go into solution and the subsequent raindrops provide good mixing turbulence and a continuously replenished "clean solvent" and (2) the particulate matter from sand size to colloidal size is dislodged by the impact of falling raindrops. Once dislodged, the particulates may be maintained in a state of pseudo-suspension by turbulence created by additional raindrops. As Sartor and Boyd suggest, qualitatively these processes may be easily described, but a further quantitative definition considering all the variables involved is a much more difficult, if not impossible, task.

Overton and Meadows (1974) cite a study prepared for the U.S. Environmental Protection Agency by Amy et al. that gives the percent of pollutant removal by street surface runoff (Table 3.4). It can be seen that the rate of removal is greatest for the first few increments of time, and thereafter rates decrease as time increases (removal is asymptotic). Runoff flowing over grass, asphalt or other urban surfaces meets with varying hydraulic resistances created by the character of the surface. Thus, sheet flow moves erratically. As it travels, it may mix with other flows having various water quality constituent concentrations; they mix and interact, and eventually they arrive at a watercourse.

Table 3.4. Percent of pollutants removed from street surfaces by runoff rate/duration.

Runoff Rate (in./hr)	Runoff Duration (hr)							
	0.25	0.5	1.0	2.0	3.0	4.0	5.0	6.0
0.1	10.9	20.5	36.0	60.1	74.9	84.1	90.0	90.0
0.2	20.5	36.9	60.1	84.1	>90.0	>90.0	>90.0	
0.3	29.1	49.8	74.9	>90.0				
0.4	36.9	60.1	84.1					
0.5	43.7	68.3	90.0					
0.6	49.8	74.8	>90.0					
0.7	55.3	80.0						
0.8	60.1	84.1						
0.9	64.5	87.4						
1.0	68.3	90.0						

Source: Overton and Meadows (1976), p. 318.

The next several sections in this chapter will present case studies in order to describe the impacts on stream water quality following the introduction of urban runoff. In the following few paragraphs, we will discuss some commonly found and well-accepted reactions which may be expected to occur from urban runoff.

High BOD loads created by urban runoff may be demonstrated in data collected by Digiano et al. (1975). They gathered BOD as well as other data at six sampling stations along the Green River in Western Massachusetts. Figure 3.8 shows the dry and wet weather results (stations progress in a downstream order from left to right). The stream winds through a rural New England landscape along Stations 1 and 2, and begins to drain urbanized areas at Sampling Station 3. Wet weather BOD concentrations are considerably higher in the latter four stations, which illustrates the effect of litter concentrations on BOD loadings. In a similar manner, wet weather also was found to increase total organic carbon (TOC), total phosphorus (P), turbidity and oil and grease.

The composition of urban litter was presented in Section 3.3. In general, as the litter accumulates, so does its eventual impact on the receiving watercourse. One of the deciding factors in the impact is the amount of time since the last rainfall. Other factors are:

(1) The intensity and the duration of the rainfall
(2) The slopes and soils within the urban area
(3) The volume and concentrations of the various biochemical constituents in the litter
(4) The interactions created by the chemical constituents washed out of the atmosphere and onto the ground by the rainfall

(5) The season
(6) The climate of the region
(7) Stream water quality at the time of the storm runoff

A phenomenon associated with the accumulation of litter and its flushing into a stream is called the "first flush effect." This is demonstrated in data collected by Weibel et al. (1964), and is shown in Table 3.5. Concentrations for the various constituents are highest immediately after the start of the rainfall, and in general tend to decrease with time. This flush effect, in most cases, seriously shock loads the receiving system.

Several studies (Bryan, 1972; Weibel et al., 1964; Brownlee et al., 1970; Federal Water Quality Administration, 1970) have shown that urban runoff contributed biochemical oxygen demands at rates which equalled if not exceeded those of the operating secondary wastewater treatment plants within the respective cities. Secondary effluent from these plants has much lower BOD than raw sewage. However, some first flushes may create shock loadings which may approach the BOD of raw sewage.

Singh et al. (1979) reported that water quality management programs in metropolitan areas are considering control and treatment of storm water and combined sewer outflow in addition to point source treatment.

3.5 Impacts of Urban Runoff on Stream Water Quality—Case Studies

The relationship between urban runoff and water quality is by no means completely known. This is a result of the fact that each urban region is unique

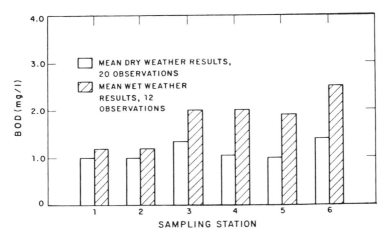

Figure 3.8. Mean dry and wet weather BOD [*Source*: Digiano et al. *Urbanization and Water Quality Control* (1975), American Water Resources Assn., Minneapolis, Minnesota].

Table 3.5. Mean concentrations of constituents in urban land runoff vs. time.

Parameter	Time After Start of Runoff				
	0–15 min	15–30 min	30–60 min	60–120 min	120 min and over
	mg/l				
SS	390	280	190	200	160
VSS	98	69	47	58	38
COD	170	130	110	97	72
BOD	28	26	23	20	12
Total Nitrogen—N	3.6	3.4	3.1	2.7	2.3
PO_4 (Total soluble as PO_4)	0.99	0.86	0.92	0.83	0.63

Source: Weibel et al. (1964), p. 923.

physically and culturally, and produces its own characteristic runoff. Therefore, the approach in the following section is to present a brief account of recent urban water quality research in various parts of the U.S. Thus, we will summarize a few selected papers, which, when taken together, will provide a sound basis for a preliminary understanding of the relationship between urban runoff and water quality.

DURHAM AND RALEIGH, NORTH CAROLINA

Rimer, Nissen and Reynolds (1978) conducted a nonpoint source assessment study to relate storm water runoff to land use. Within this study area (about 2000 square miles, or 5200 square kilometers) 11 automatic samplers were set up at selected land use locations. Water stage recorders were also established at nine of the sites, and the network's five rain gauges were increased by six. In addition to the sampling sites, six continuous water quality monitors were located throughout the drainage areas. These units measured specific conductance, pH, dissolved oxygen and temperature. Land uses were classified into seven categories, which ran the gamut from (1) low activity rural to (7) urban—predominately central business district.

Seven automatic sampling stations were located on streams draining a single predominant land use. These were relatively small areas, and the streams had very low baseflows. Whenever storm runoff occurs, the vast majority of the streamflow is from surface runoff from a single predominant land use.

A storm occurred on September 22, 1976, and Figures 3.9, 3.10 and 3.11 show the pollutographs from three selected stations. This was a relatively short-duration storm with peak intensities of up to about 1.2 in./hr (3.0 cm/hr). There was no recorded rainfall for three days prior to the event.

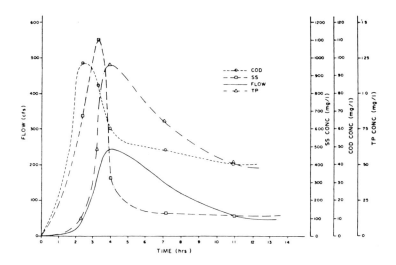

S-1 SR 1730. Pollutograph analysis samples taken Sept. 22–23, 1976.

Figure 3.9. Pollutograph of a rural area [*Source*: Rimer et al. (1978) p. 256].

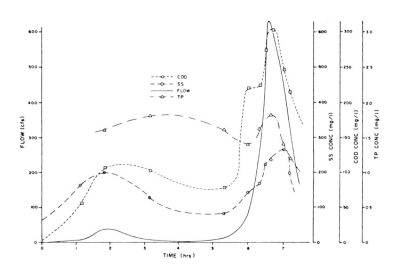

S-11 W. Club Blvd. Pollutograph analysis samples taken Sept. 22, 1976.

Figure 3.10. Pollutograph of a high activity residential area [*Source*: Rimer et al. (1978) p. 257].

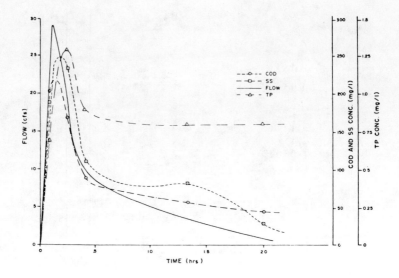

S-8 Cabarrus St. Pollutograph analysis samples taken Sept. 22, 1976.

Figure 3.11. Pollutograph of a central business district [*Source*: Rimer et al. (1978) p. 258].

Figure 3.9 shows that suspended solids and phosphorus levels increase significantly as does COD concentration. Also note that the pollutograph peaks precede that of the discharge, indicating the effect of the first flush. Figure 3.10 illustrates data from a high-activity residential area (note that the time axes and concentration axes are different for the first three figures). Suspended solids rose substantially, but not as high as the rural stations. Phosphorus concentration and COD concentrations exceeded those of the first station. The effects of the increased impervious area are demonstrated by the close tracking of the four curves.

Figure 3.11 demonstrates data collected from a central business district. Notice the rapid risetime for all four curves, all peaking within a 10-min period. Total phosphorus levels were comparable to the rural station and COD concentrations were above those of the other stations. Suspended solids concentrations were below those of the rural and residential areas.

MILWAUKEE, WISCONSIN

Cherkauer (1975) selected two drainage basins, one rural and one urban, for study. Both basins were physiographically and climatically similar, located a few miles from each other (their drainage areas do not touch). The urban basin had a drainage area of 2.9 mi^2 (7.5 km^2) and the rural basin had an area of 3.75 mi^2 (9.7 km^2). Table 3.6 lists the pertinent land use data.

Water samples were collected from both streams before, during and after the rainstorm of 0.87 in. (2.2 cm) which occurred on October 6, 1974. Rainfall was monitored at one site in each of the study areas. No recording stage gauges were available, and all discharge measurements were made manually at a greater frequency than the water sampling.

Cherkauer points out that prior to the storm, the urban basin had a high baseflow. He attributes this to the higher percentage of impermeable ground cover, and storm sewer input (water from lawn waterings, street cleaning). After seven days with no rain, a 0.87-in. (2.2-cm) storm occurred and the peak flow observed from the urban basin was over 250 times that of the rural basin (Figure 3.12).

Figure 3.13 shows that at baseflow, the rural stream has the highest concentrations of dissolved constituents. The pH was lower and the water temperature higher. Cherkauer points out that these observations result from the fact that the urban stream had a measurable baseflow, whereas the rural stream's baseflow was virtually stagnant. The rural stream had high dissolved solids concentrations because of evaporation and the pH was lower because of the higher temperature which enhanced the solubility for dissolving CO_2.

Surface runoff from the storm resulted in a dilution of all dissolved constituents in both streams (Figure 3.13). Cherkauer defines the degree of dilution by the following formula:

$$\text{dilution percentage} = \frac{\text{baseflow concentration} - \text{peak flow concentration}}{\text{baseflow concentration}}$$

Figure 3.14 shows that the urban dilution was greater than that in the rural stream. Cherkauer points out that this means that urban surface runoff for this

Table 3.6. Land use of the drainage basin.

Use	Brown Deer (Urban) Total Area = 7.5 km²		Trinity (Rural) Total Area = 9.7 km²	
	Percent of Basin Area	Subtotals	Percent of Basin Area	Subtotals
Residential	56.9		5.0	
Light Industry	5.7		0.3	
Parking Lots	2.5		0.4	
Total Developed		65.1		5.7
Open Land	33.4		94.3	
Construction Sites	1.5		0.0	
Total Undeveloped		34.9		94.3

Source: Cherkauer (1975), p. 988. Water Resources Bulletin, Vol. II, No. 5, pp. 987–998. The American Water Resources Assn., Minneapolis, Minnesota.

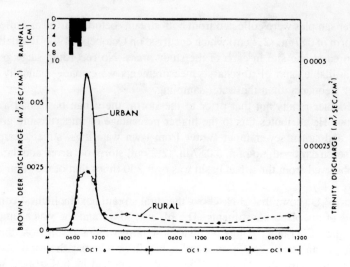

Figure 3.12. Discharge per unit drainage area hydrographs and rainfall histogram [*Source*: Cherkauer (1975) p. 990. *Water Resources Bulletin*, Vol. II, No. 5, pp. 987–998. The American Water Resources Assn., Minneapolis, Minnesota].

particular storm had a lower dissolved ion concentration than the runoff from the rural basin. Since the urban storm water ran off faster, there was a shorter contact time between the runoff and the natural materials. This produced the interesting paradox that during the high water event, the water quality of the urban stream was better (in terms of dissolved solids) than that in the rural stream.

Cherkauer points out that the urban stream, despite the lower ion concentrations, carried out vastly greater quantities of dissolved material (Table 3.7).

Table 3.7. Total loads of various materials per unit drainage area carried out of basin between 0530, October 6, and 0900, October 8, 1974.

Material	Urban Load (kg/km^2)	Rural Load (kg/km^2)
Water	86,300	675
Suspended Sediment	287	0.8
Total Dissolved Solids	199	5.6
Sodium	25.7	0.6
Chloride	37.5	0.7
Calcium	23.7	1.0
Magnesium	11.6	0.2
Bicarbonate	61.3	3.4

Source: Cherkauer (1975), p. 994. *Water Resources Bulletin*, Vol. 11, No. 5, pp. 987–998. The American Water Resources Assn., Minneapolis, Minnesota.

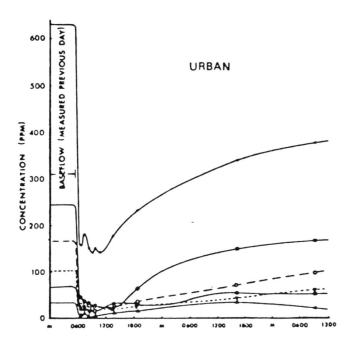
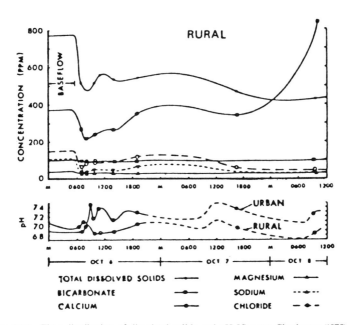

Figure 3.13. Time distribution of dissolved solids and pH [*Source*: Cherkauer (1975) p. 991. *Water Resources Bulletin*, Vol. II, No. 5, pp. 987–998. The American Water Resources Association, Minneapolis, Minnesota].

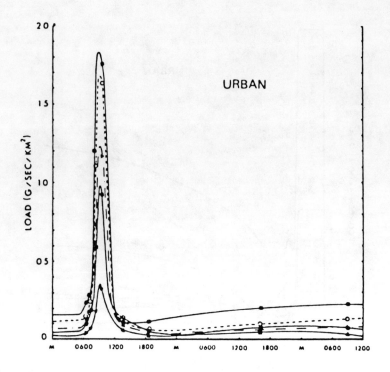
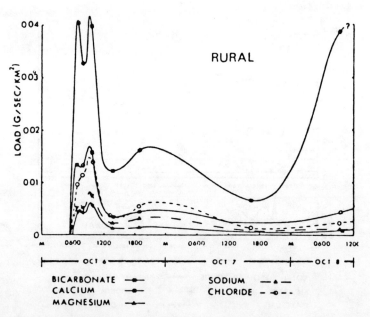

Figure 3.14. Time distribution of loads per unit area of various dissolved solids during the flood [*Source*: Cherkauer (1975) p. 993. *Water Resources Bulletin*, Vol. II, No. 5, pp. 987–998. The American Water Resources Assn., Minneapolis, Minnesota].

DURHAM, NORTH CAROLINA

Bryan (1972) conducted a study of a 1067-ac (4.3-km^2) watershed to determine the yield of pollutants. Within this area was a portion of the central business district of Durham, a tobacco processing and manufacturing plant, a cemetery, churches, scattered commercial business establishments, public schools, extensive open space for recreation, a shopping center and high- and low-population density housing districts. An expressway within the watershed was in the final stages of completion during the study period. The U.S. Geological Survey maintained a stream gauging station and a rain gauge within the study area. About 60% of the basin was residential, about 20% devoted to commercial and industrial activities, and about 20% to public, unused or institutional categories. Water samples were manually collected during 30 storms occurring over a 14-month period. Seventeen storms were sampled more thoroughly, and were judged to yield the most reliable results.

Table 3.8 lists the summary of data collected from the study area. In a manner similar to Weibel et al. (1964), Bryan points out the fact that his data show a pollutant capability similar to that of a secondary wastewater treatment plant. Table 3.9 offers a meaningful comparison of the Durham data to data collected by other researchers.

Bryan concluded that the estimated annual contribution of BOD by storm water was equal to that of the drainage area's sanitary wastewater effluent from secondary treatment, and the COD was greater than the discharge of raw, sanitary sewage from strictly residential, average urban area of the same size.

CINCINNATI, OHIO

Weibel, Anderson and Woodward (1964) studied a 27-ac (0.11-km^2) residential and light commercial drainage basin with separate sewers in the Mt.

Table 3.8. Estimate of pollutant yields from the urban drainage basin in Durham, North Carolina.

Parameter	Annual Yield (lb)	Annual Yield (lb/ac)	Annual Yield (lb/mi^2)
BOD	90,000	84	54,000
COD	1,110,000	1,040	665,000
TS	16,900,000	15,900	10,140,000
VTS	1,850,000	1,730	1,100,000
NaCl	78,200	73	46,800
PO$_4$	3,600	3.4	2,160
Pb	2,000	1.9	1,190

Source: Bryan (1972), p. 584. Water Resources Bulletin, Vol. 8, No. 3, pp. 578–588. American Water Resources Assn., Minneapolis, Minnesota.

Table 3.9. Comparison of water pollutants in Durham and other cities.

Location	Parameter	BOD (mg/l)	COD (mg/l)	Total Solids (mg/l)	Volatile Solids (mg/l)	Suspended Solids (mg/l)	Total Phosphate (mg/l)	Fecal Coliforms (per 100 ml)	Chloride as NaCl (mg/l)
Durham, NC (urban storm water)	Mean	14.5	179	2730	298	—	0.58	30,000	12.6
Cincinnati, OH (urban storm water)	Range	2–>232	40–600	274–13,800	20–1110	—	0.15–2.50	7000–86,000[a]	3.0–390
	Mean	17	111	—	—	227	1.1	—	19.8
Cincinnati, OH (rainfall)	Range	1–173	20–610	—	—	5–1200	<0.02–7.3	500–76,000	5.0–705
	Mean	—	16	—	—	13	0.24	—	—
Coshocton, OH (rural storm water)	Mean	7	79	—	—	313	1.7	—	—
Coshocton, OH (rainfall)	Range	0.5–23	30–159	—	—	5–2074	0.25–3.0	<2–56,000	—
	Mean	—	9.0	—	—	11.7	0.08	—	—

Table 3.9. (continued).

Parameter / Location		BOD (mg/l)	COD (mg/l)	Total Solids (mg/l)	Volatile Solids (mg/l)	Suspended Solids (mg/l)	Total Phosphate (mg/l)	Fecal Coliforms (per 100 ml)	Chloride as NaCl (mg/l)
								Total Coliforms MPN/100 ml	
Detroit, MI (1949) (urban storm water)	Range	96–234	—	310–914	—	—	—	25,000–930,000	
Seattle, WA (urban storm water)		10	—	—	—	—	4.3 max	16,100 max	
Stockholm, Sweden (urban storm water)	Median	17	188	300	90	—	—	4000	
	Maximum	80	3100	3000	580	—	—	200,000	
Pretoria, S. Africa									
(residential/park/school)		30	29	—	—	—	—	240,000	
(business and flat area)		34	28	—	—	—	—	230,000	
Oxney, England	Maximum	100	—	—	—	2045	—	—	
Leningrad, USSR		36	—	—	—	14,541	—	—	
Moscow, USSR	Range	18–285	—	—	—	1000–3500	—	—	

[a]Range of means for 17-storm series for Durham, NC.
Source: Bryan (1972), p. 585. Water Resources Bulletin, Vol. 8, No. 3, pp. 578–588. The American Water Resources Assn., Minneapolis, Minnesota.

Washington section. The population density of this region was nine persons per acre (2224 persons/km^2). Impermeable area accounted for 37% of the watershed, which consisted of roofs, paved streets and paved parking lots. The remainder was made up of lawns, gardens and parks.

The study area was instrumented with a rain gauge, and storm water flows were measured with a 4-ft (1.22-m) rectangular weir and a record was maintained by a continuous water level recorder. Water samples were collected by an automatic sample. A float device located in a sewer actuated the pump whenever a flow of water occurred. The pump would then fill special plastic sampling bottles. In this manner, data was collected from July, 1962 to September, 1963, exclusive of January and February.

Table 3.10 shows the means and ranges of the various parameters collected. The average BOD and COD was about at the level of secondary sewage treatment plant effluent. Suspended solids average was 210 mg/l, which is about the concentration found in raw sewage. Volatile suspended solids (VSS) concentration exceeded that of sewage treatment plant effluent. The nutrients, i.e., nitrogens and phosphates, were in high concentrations, and turbidity and color were also of consequence. Pesticides were also found in considerable quantities.

Table 3.11 lists the bacterial counts of samples collected. Fecal coliform counts are lower than the others, which, as pointed out by the authors, probably indicates predominantly nonhuman sources of pollution (in this instance, dogs and cats).

This study was one of the first completed, and its results alerted scientists to the pollutant capabilities of urban runoff. Even though data gathered in this study could not be compared immediately to those from other urban areas, they nevertheless gained considerable impact when they were compared against computed sanitary sewage discharge data from the same area. These data are shown in Table 3.12 and clearly illustrate the pollutant capabilities of urban runoff.

3.6 Impact of Urban Runoff in Erosion

Sediment is defined in a geologic sense by Colby (1963) as:

> . . . fragmental material that originates from the disintegration of rocks and is transported by, suspended in, or deposited by water or air, or is accumulated in beds by other natural agencies (p. vi).

Hawkinson, Ficke and Saindon (1977) add to this definition when they write:

> . . . sediment in streams results from erosion of soils by overland (sheet) erosion, and by scouring of ditches and stream channels. Flowing streams carry some sediment almost all of the time, but by far the highest concentrations and biggest loads are carried by storm runoff.

Table 3.10. Constituent concentrations in urban land runoff at a Cincinnati sampling point, July 1962 through September 1963.[a]

Parameter		Range	Mean
		(units)	
Turbidity		30–1,000	170
Color		10–380	81
pH		5.3–8.7	7.5
		(mg/l)	
Alkalinity		10–210	59
Hardness (as $CaCO_3$)	Ca	24–200	63
	Mg	2–46	15
	Total	29–240	78
Cl^-		3–35	12
SS		5–1,200	210
VSS		1–290	53
COD		20–610	99
BOD		2–84	19
Nitrogen (as N)	NO_2	0.02–0.2	0.05
	NO_3	0.1–1.5	0.4
	NH_3	0.1–1.9	0.6
	Org.	0.2–4.8	1.7
PO_4 (total soluble as PO_4)		0.07–4.3	0.8

[a]January and February not included.
Source: Weibel et al. (1964), p. 919.

Table 3.11. Bacterial counts in storm water runoff samples from an urban area.[a]

	Counts Exceeded in Designated Percent of Samples (Colonies per 100 ml)		
	90%	50%	10%
Coliforms[b]	2,900	58,000	460,000
Fecal coli[b]	500	10,900	76,000
Fecal strep[c]	4,900	20,500	110,000

[a]Samples were taken from the Cincinnati study site during 1962 and 1963.
[b]Coiforms and fecal coliforms were determined by the completed MPN technique during 1962, and by the MF technique during 1963.
[c]Fecal streptococci were detected by KF streptococcus agar pour plates.
Source: Weibel et al. (1964), p. 920.

Table 3.12. Comparison of storm water runoff loads and sanitary sewage loads (lb/yr/acre).[a]

Constituent	Storm Water Runoff[b]	Sanitary Sewage (raw)[c]	% Storm/Sanitary
SS	730	540	140
VSS	160	360	44
COD	240	960	25
BOD	33	540	6
PO$_4$	2.5	27	9
Total nitrogen—N	8.9	81	11

[a]Multiply by 1.12 to obtain kg/yr/hectare.
[b]Storm water runoff: based on essentially complete measurement of rainfall and consequent runoff water quantity and quality at the study site during September through November 1962, and March through September 1963, projected to average annual rainfall at Cincinnati.
[c]Sanitary sewage: based on population density of 9 persons/acre (22/hectare) and flow of 100 gpd/cap (0.4 m^3/day/cap).
Assumed raw sewage strengths: SS—200 mg/l; VSS—130 mg/l; COD—350 mg/l; BOD—200 mg/l; PO$_4$—10 mg/l; total N—30 mg/l.
Source: Weiber (1964), p. 923.

Erosion is a natural process; the earth is continuously being eroded. This is part of the natural scheme of things. However, within urban areas, increases in storm runoff add high peaks of energy which augment the natural erosive forces and greatly accelerate erosion.[23] Any unprotected ground surface may easily be scoured. Streams are filled with sediment-laden water, and their cross-sectional areas may be enlarged (Hammer, 1972; Dawdy, 1967). In contrast to the most invisible character of chemical and biochemical constituents of urban runoff, suspended sediment is easily observable and has been the subject of much concern. As Wolman and Schick (1967) note:

> ... imposition of large quantities of sediment on streams previously carrying relatively small quantities of primarily suspended materials produces a variety of changes in the physical and biological characteristics of the stream channel. These changes include deposition of channel bars, coarsening of suspended sediments in the channel, erosion of channel banks as a result of deposition within the channel, obstruction of flow and increased flooding, shifting configurations of the channel bottom, blanketing of bottom dwelling flora and fauna, alteration of the flora and fauna as a result of changes of light transmission and abrasive effects of sediment, and alteration of the species of fish as a result of changes produced in the flora and fauna upon which the fish depend (p. 458).

From this observation, one can clearly understand why suspended sediment may be termed a pollutant (also see Swerdon and Kountz, 1973). In large enough quantities, it can seriously disrupt a stream's ecology.

[23]See Section 7.3 for a discussion of sheet erosion.

Suspended sediment concentrations measured in urban streams may vary greatly depending on many physiographic and climatic factors. It is obvious that if a surface is stripped bare of vegetation and is exposed to rainfall and urban runoff waters flowing from impervious surfaces, very high suspended sediment concentrations will result. Urban areas contain large amounts of impervious surfaces, as has been shown; however, they do not necessarily contain easily erodible surfaces. Therefore, high suspended sediment concentrations are probably indicative of: (1) land surface disturbance somewhere within the watershed, (2) accumulation of dust and the first flush effect, and (3) scouring of the stream channel itself. The first factor is the topic of discussion for the two case studies below. The second factor, the first flush, may be determined by the lagtime of the pollutograph. If the lagtime is relatively long, then the actual weight of eroded materials is probably not of much consequence. Scour of the stream channel has been proposed by several researchers as being indicative of urban erosion. Hammer (1971) reported the effects of urbanization on stream channel enlargement. Essentially, the increase in peak flows brought about by urban land use causes scouring of stream channels. If a researcher is collecting suspended sediment data during one of these peaks, high concentrations may be observed.

Several researchers have shown that without a doubt, nowhere in the process of urbanization is erosion so violent as in the construction phase.[24] Guy (1963) was perhaps the first researcher to point out the tremendous increase in sediment discharge occurring downstream of an area undergoing residential construction. Wolman and Schick (1967) indicated that average sediment yield for a representative selection of large and small rural or wooded drainage areas in Maryland would be on the order of 200–500 ton/mi^2/yr (181–454 metric tons/km^2/yr). Stable urban areas could also fall in this category as suggested by Wolman. In the same paper, Wolman and Schick point out that small areas undergoing construction could yield up to 140,000 tons/mi^2/yr. Graf (1975) reported that 7544 metric tons/mi^2/yr (8315.7 tons) were deposited on new flood plains in a small suburban Denver watershed following construction. This represented an increase of 30 times in sediment production over the preconstruction rate. Not only is the receiving stream structure affected (Wolman, 1967), but also, as pointed out above, the stream ecology is upset and there is tremendous loss of topsoil, a valuable and nonrenewable resource.

MONTGOMERY COUNTY, MARYLAND

Yorke and Herb (1978) completed a study of streamflow and suspended sediment from nine drainage sub-basins within a 32-mi^2 area between the

---

[24]Dawdy (1967) calls this the "transient yield" and points out that this is an unpredictable and unmeasured relation.

years 1962 and 1974. They employed data from 9 streamflow-sediment stations, 9 recording rain gauges, numerous nonrecording rain gauges, 14 sets of aerial photographs, and 3 channel surveys along stream reaches. Three sub-basins remained rural during the study period, while the others underwent urban development. In 1974, urban area occupied 0–60% of each sub-basin.

One of the basins accounting for 21.1 mi^2 (Northwest Branch, Anacostia River, near Colesville) had an average annual sediment yield of 14,800 tons (13,427 metric tons) during the study period. This ranged from 5500 tons (4990 metric tons) in 1974 to 31,000 tons (28,123 metric tons) in 1972; (the latter was greatly influenced by hurricane Agnes). The majority of the sediment loads were transported during storms: 73% in 2.2% of the time, and 94% in 5.7% of the time. Sediment yields were highly variable from storm to storm, and loads were most closely related to the volume of storm runoff and peak discharge.

The major sources of sediment were cropland, urban land and construction sites. Yields of suspended sediment from lands under cultivation ranged from 0.65 to 4.3 tons/ac (145.8 to 963 metric tons/km^2), and forest and grasslands were estimated to produce 0.03 to 0.2 tons/ac (6.67 to 44.5 metric tons/km^2), respectively. Yields from urban land were about 3.7 tons/ac (830 metric tons/km^2), mostly from stream-channel erosion immediately downstream of the newly completed residential and commercial areas.

Urban construction sites produced the largest annual yields of suspended sediment, with a range from 7 to 100 tons/ac (1569 to 22,416 metric tons/km^2) and an average of 33 tons/ac (7397 metric tons/km^2).

Highway Construction

During periods of highway construction, erosion may occur in a similar, although usually less dramatic, manner to that of urban construction. Dawdy (1967) listed research that reported yields of up to 3000 tons/mi^2 (1051 metric tons/km^2) from highways in the piedmont of Maryland. Since many of these highways are not only built in urban areas, but also constructed to aid interurban communication and development, they are of importance to urban hydrology.

SCOTT RUN BASIN, FAIRFAX, VIRGINIA

Vice, Guy and Ferguson (1969) studied sediment movement as caused by suburban highway construction in a 4.54-mi^2 (11.76-km^2) watershed during the years from 1961 through 1964. Highway construction covered 11% of the basin, and other types of development during this period were minor. Examination of aerial photographs flown in August, 1960 and September, 1964, revealed that land in forest and trees occupied 50% of the basin in 1960, and decreased to 43% by 1964. The area was predominantly rural prior to highway construction. Farming operations occupied 25-30% of the basin area; however, only a small part of the farmland was used for cultivation at any one time. The major part of the farmland was grass cover, used for hay and grazing. Residential and industrial areas occupied about 10% of the basin in January, 1961 and about 12% in October, 1964. The authors point out that this area produced a low yield of sediment since most of the drainage surface was protected by pavement, roof or sod.

A continuous record of stream discharge (when flow was above 20 cfs) was provided by a gauging station on Scotts Run, and precipitation data was obtained from three nearby weather bureau stations. Sediment transport data were computed for twenty-nine storms.

Five interchanges and three dual highways were constructed across the basin during 1960–1962. The peak construction period occurred during the winter of 1961–62 and was terminated by fall of 1962. In 1964, a small length of highway was constructed. The authors report that residual erosion effects declined rapidly in the fall of 1962 and spring of 1963, but some effect persisted to the end of the study period in September 1964.

The authors, following methods adapted from the research of others, divided the basin into low-yield, intermediate-yield and high-yield construction areas. A low-yield area would be expected to produce 0.03 tons/ac/yr (6.72 metric tons/km^2/yr), and would be stable forest area in this region. Grass and established urban areas were predicted to have an average yield of 0.25 tons/ac/yr (56.0 metric tons/km^2/yr). Cultivated fields or areas of intermediate yield would be about 13 tons/ac/yr (2914 metric tons/km^2/yr).

Construction area sediment yields vary considerably from region to region for various practical and physiographic reasons. At any one time, highway construction areas occupied from one to more than ten percent of the study area watershed. The authors computed that during the study period, 33,320 tons (30,228 metric tons) of sediment, or 85% of the total, passed the gauging station. The low-yield forest areas contributed 0.3%, grass and established urban, 3.0%. The intermediate-yield areas contributed 4.1%, and other kinds of construction contributed 7.5%. The highway construction area produced an average annual yield of 63 tons/ac (14,122 metric tons/km^2).

Examination of the drainage area total sediment discharge during the study period shows that the 88 storms carried 94% of the total sediment load. Five percent of the sediment discharge was made up of bedload, which comprises the larger particles which are dragged along the bed of the stream and sometimes, depending on flow, may bounce up and be carried in suspension. The remaining 1% of the sediment total was carried by low flow discharges. In other words, the 88 events occupied 3% of the time, but carried 94% of the sediment discharge for the period.

The authors point out that precipitation during the study period was about

14% less than the long-term average and, had the mean precipitation prevailed, a considerable increase in sediment yields would have occurred.

3.7 Impact of Urban Runoff on Stream Water Quality

ROAD SALT

During winter periods since 1940 many highway authorities have adopted a "bare pavement" policy. This has been accomplished by spreading deicing salts on highways and streets. Within urban areas, the density of streets and the need to keep a large volume of traffic safely moving have created a considerable concentration of salts. These salts remain within the hydrologic environment well after the end of the winter season. Recently, several studies have been completed which have shown increasing chloride concentrations in surface and subsurface waters, and have alerted scientists to their possible environmental impact.

Wulkowicz and Saleem (1974) studied the flow of road salts through a 377-mi^2 basin in Northeastern Illinois for the period from November 1972 to April 1973. The method they employed was to calculate the material balances, i.e., to compute an annual salt chloride budget. Population of the thirty major communities within the basin increased 46% in the decade between 1960 and 1970 (Spieker, 1970). Street mileage increased 14% in the same decade. Most of this growth was in residential streets.

From eleven separate locations, a total of 612 water samples were collected. Whenever possible, samples were collected during anticipated periods of high runoff, usually biweekly. At the time of the study, there were sixteen sewage treatment plants discharging a total of 2.1 m^3/sec into the stream. Since chloride is a constituent of these discharges (humans use salt with almost every meal), it had to be accounted for in the final budget.

Of the forty-one road maintenance agencies, thirty-six reported their salt usage totals to the authors. This accounted for an estimated 97.2% of the road salt total used during the period. A total of 20,260 tons (18,380 metric tons) of NaCl and 40 tons (36.29 metric tons) of CaCl$_2$ was applied to the 6500 lane kilometers (4040 miles) of streets within the basin during the study period.

The authors use the following formulas:

$$L = CQ$$

where

L = chloride load at any time at a location along a stream
C = the concentration of chloride in water
Q = water discharge

The road salt component L_s of the total load is given by:

$$L_s = L_o - L_c$$

where

L_o = the observed total load
L_c = the load due to all nonroad salt sources

L_c must be computed from the background chloride concentrations, and was determined from the chloride-discharge relations for each stream. A portion of L_s (road salt chloride load) is removed from the watershed by the stream; the remainder is either retained or is removed by the groundwater.

The Salt Creek Basin was divided into four subsections; certain characteristics of each are listed in Table 3.13. An isochlor diagram was developed by Wulkowicz and Saleem in an attempt to visually portray the changes in chloride concentration with distance downstream (Figure 3.15) versus the time in the study period. December and February each show periods of major water quality impairment; March shows a minor period. Each one of these periods corresponds to recorded periods of major salt application. Dilution is greatest at the center section of the stream, since two minor tributaries (which drain areas of at least 65% open space with low highway and population density) join the main stream at this point. The upstream reach has high chloride concentrations because of the funneling of runoff from three communities clustered in the area of Stations 9, 10 and 11. The bulk of the population is located in the downstream reaches and road salt use is heaviest here, as evidenced by the high concentrations.

The authors point out that each major water quality impairment period has double maximums of chloride concentrations. The first is the salt-charged runoff produced by the salt-induced initial melting of snow. The interim low is caused by the dilution of the road salts due to the melting of snow induced by them, and subsequently by a reduction in runoff with the advent of temperatures below $-10\,°C$ (NaCl loses effectiveness below $-10\,°C$). After the storm has subsided, thawing conditions melt the remaining snow and flush the residual salts into the stream and produce the second maximum.

After each episode, stream chloride concentrations decline. This occurs due to dilution by runoff of precipitation and melting snow. Decline rate is affected by the intensity of precipitation as well as by the length of the thawing period after major salting applications. A four-year recurrence interval flood in December created a sharp decline in chloride values. A gradual decline was experienced in February as a result of the normal thawing trend aided by short duration rainfalls in March.

During the six-month study period, the authors computed a removal of the

Table 3.13. Urban aspects of the Salt Creek Basin.

Urbanization Parameters	Subsection I	Subsection II	Subsection III	Subsection IV	Basin	Correlation Coefficient[a]
Drainage area, km²	76.9	205.9	44.6	49.2	376.5	
Area as streets,[b] %	8.7	11.5	16.8	18.1	12.3	0.928
Highway density, lane km/km²	14.2	17.4	16.3	22.7	17.5	0.934
Population[c]	49,400	166,200	44,300	98,100	358,000	
Population density, people/km²	642	807	993	1,994	951	0.936
Chloride t/km²	30.6	30.6	33.3	41.2	32.4	0.909
Chloride removed %	55.5	61.4	64.5	71.9	62.4	

[a] Plotted against the percent of chloride removed.
[b] From the Northeastern Illinois Planning Commission (unpublished data, 1970).
[c] Population values are for the 30 major communities in the basin and are not total population for subsections. Data are from 1970 Census.
Source: Wulkowicz and Saleem, p. 980. *Water Resources Res.*, 10(5):974–982 (1974), copyrighted by American Geophysical Union.

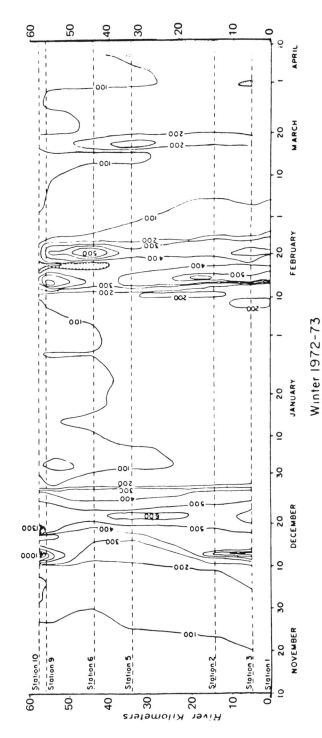

Figure 3.15. Isochlor diagram based on the observed chloride concentrations in water of the Salt Creek basin [*Source:* Wulkowicz and Saleem, p. 977. *Water Resources Res.* 10(5):974–982 (1974), copyrighted by American Geophysical Union].

applied salt of between about 55 and 72% for the various subsections. The areawide average was 62.4% (Table 3.14). When the degree of urbanization for each subsection was plotted against amount of chloride removed, a correlation coefficient above 0.9 was obtained. This agrees with conclusions presented by the American Public Works Association (1969) in that virtually 100% of the applied road salt chloride can be expected to wash into storm drains from impervious surfaces.

A summary of the road salt budgets is shown in Table 3.15. It is interesting to note that the northern part of the basin had the highest concentrations of chloride, yet removed the least amount of road salt. The authors state that this is a result of that part of the creek having limited dilution capacity.

The authors estimate that 10% of the remaining 37% of the road salt chloride was removed by surface flow in the six-month period following the study period. The remaining portions were either removed by groundwater outflow, or retained in the basin because of lag mechanisms.

LEAD

Newton, Shephard and Coleman (1974) observed ". . . lead is a primary contaminant of automobile emissions and tends to be deposited within a short distance of the roadbed" (p. 999). Since there is a large concentration of automobiles in urban areas, lead accumulates in considerable quantities. Automobiles emit lead at a rate which is approximately 0.11 g/mi (0.07 g/km). The authors compute that 143,000 g/day (315 lb) of lead is emitted by cars in Oklahoma City, and they speculate that as much as 50% finds its way to streams (lead poorly adsorbs to road surfaces, but readily adsorbs to soils). Lead is toxic to fish as well as toxic to humans. The continuous concentration of lead in urban areas and its flow into streams is becoming an increasingly more serious issue.

Ottawa, Ontario

Lead accumulation in urban snow was studied by Oliver, Milne and LaBarre (1974). They collected snow samples weekly from eleven snow dumping sites and periodically from selected highways and streets throughout the city. Water samples were collected weekly from Ottawa storm sewers and daily from February 21 to March 24, 1972, from the wastewater treatment plants. The population of Ottawa with suburbs is about 500,000. The authors state that 160,000 vehicles enter the core of the city daily.

Lead levels in snow along the city roads were found to correspond roughly to traffic volume. The mean level of lead in snow from the disposal sites was 4.8 mg/l. Lead levels in each disposal site tended to be diverse, since they reflected the lead concentrations occurring at the streets from which the snow

Table 3.14. Annual road salt chloride budgets for different areas of the United States.

Study Area	Year	Salt Used (t)	Chloride Removed (%)	Reference
Salt Creek Basin, Illinois[a]	1972–1973	20,260	62.4	This study
Irondequoit Basin, New York	1971–1972	68,900	68.0 (44.)	Diment et al. (1973)
Irondequoit Basin, New York	1970–1971	72,900	48.0 (37.)	Diment et al. (1973)
Irondequoit Basin, New York	1969–1970	77,000	41.0 (29.)	Bubeck et al. (1971)
Boston area	1969–1970	15,100	65.0	Huling and Hollocher (1972)
Sleeper's River Basin, Vermont	1969–1970	67–100	60.–89.[b]	Kunkle (1971)
Chicago, Illinois[c]	1967–1968	46.1	100.0	American Public Works Association (1969)

[a]The values reported are for the first 6 months following initiation of salting.
[b]The mean was 74%.
[c]This budget is of 1.5 months duration.
Numbers in parentheses are estimated road salt chlorides removed for the period equivalent to this study based on data from the original work.
Source: Wulkowicz and Saleem, p. 980. Water Resources Res., 10(5):974–982 (1974), copyrighted by American Geophysical Union.

Table 3.15. Summary road salt chloride budgets for the Salt Creek Basin, November 1972 through April 1973.

Subsection	Observed Chloride Load (t)	Calculated Background Load (t)	Chloride Used (t)	Chloride Removed by Creek %
I	2,774	1,468	2,354	55.5
II	11,860	7,997	6,294	61.4
III	2,100	1,141	1,485	64.5
IV	1,395		1,940	71.9
Basin	18,203[a]	10,606	12,176[a]	62.4

[a]This value contains 170 t of NaCl from the excluded area of the basin.
Source: Wulkowicz and Saleem, p. 979. Water Resources Res., 10(5):974–982 (1974), copyrighted by American Geophysical Union.

was collected. Snow was analyzed from two city roofs and found to contain over ten times less lead. This indicated that the streetside snow was indeed heavily contaminated by automobile exhausts.[25]

The authors noted that most of the lead tended to associate with fine particles where the surface area for adsorption and/or chemical reaction is considerably greater. Lead seems to adsorb on the particulate matter in snows and yet not be present as compounds. The authors suggest that possibly these particles contain limestone, which seems to be a good medium for adsorption. This indicates that lead easily finds its way into rivers by suspended solids transport, and there it accumulates in sediments.

At the snow dump sites, the lead is retained after the snow melts and is concentrated. The writers estimated that 2,387,000 yd^3 of snow were dumped during the winter of 1971–1972 in Ottawa. From this, they estimated an accumulation of 10,800 lb (4.90 metric tons) of lead. If this snow had been dumped into the area watercourses, one can imagine the environmental ramifications. Fortunately, these dump sites were more than 100 ft (30.5 m) from any river, and it is estimated that the input of lead into the rivers was reduced to 2%. Lead soil profiles confirm the fact that the snow dump sites had the remaining lead. The authors calculate that 16,000 lb (3.63 metric tons) of lead were contributed to the waters by storm sewers and about 9000 lb (4.08 metric tons) were discharged in wastewater effluent in the study year. They estimate that the moving of the snow sites (to 100 ft from the watercourses) reduced the lead flow into the river by 30%.

3.8 Summary

The water quality of natural streams is related to (1) water quantity (surface and subsurface runoff) of the stream, (2) the geology through which it flows, and (3) the climatic and geologic histories of the region. These factors are interrelated and the individual effect(s) of each would be impossible to isolate.

In the short term, the water quantity/water quality-related factors give rise to the various biotic zones found along rivers. A large flood may temporarily change these zones, but in a relatively short period of time (in reference to geologic time) these biotic zones should reestablish themselves.

When runoff which has higher concentrations of constituents than normal is discharged into a stream, the water quality balance of the stream system may be upset. A stream has a certain assimilative capacity, and may dilute and ac-

---

[25]Oliver et al. point out that automobile gasoline contains between 1.5 and 3.6 g of lead per gallon (3.8 liters), in the form of tetramethyl or tetraethyl lead.

cept concentrations up to a certain level, depending on many flow and environmental factors. When the assimilative capacity of a stream is exceeded for a period of time, then a biotic zone may vary in magnitude and distribution, and the stream may be called "polluted." New flora and fauna establish themselves and may thrive (at the expense of the former inhabitants) on the new concentration levels.

Raw sewage is an excellent food source for many types of bacteria, and logarithmically stimulates their growth. In this case, a dissolved oxygen sag curve is established in a stream and may extend downstream for tens of miles, reducing the recreation potential of the river course and increasing the cost of water processing for municipalities downstream.

Research has shown that urban nonpoint source runoff may be as potent as raw sewage discharge for some constituents, but it generally has a BOD loading approximately equal to secondary sewage treatment plant effluent. Different land uses within an urban area vary in their capabilities to produce such runoff. In totality, each urban region is unique and the outcome of the meshing between water quantity and water quality is also unique.

In the study conducted by Rimer et al. (1978), the potency of urban runoff (in this case, from a highly impervious central business district) is demonstrated by Figure 3.11. In a matter of minutes, COD and suspended solids concentrations of about 270 mg/l and total phosphorus concentrations of about 1.3 mg/l were washed into the receiving watercourse. This represented a substantial loading from the first flush effect. When the time of this pollutograph is compared to those of Figures 3.9 and 3.10, one can clearly see the shock loading impact an urban area can have on the water quality of its receiving watercourse. Depending on the size of the urban area and the size of the stream, urban nonpoint source runoff can considerably upset stream water quality and can rapidly generate a dissolved oxygen sag curve.

In the study conducted by Cherkauer (1975), it is shown that dilution is greater in the urban stream than in the rural. As suggested by the author, this is a result of the faster rate of urban runoff and the consequent shorter contact time between the runoff and the natural materials. All in all, this produced the paradox (with respect to the dissolved solids) that the urban stream water quality was better than that in the rural stream. This study shows that water quality (depending on constituents investigated) may not always be deteriorated by urban runoff.

The Bryan (1972) study supported the conclusion, brought out earlier by Weibel et al. (1964) in another locale, that the BOD contribution by storm water was equal to that of secondary treatment effluent from the drainage area's sanitary wastewater treatment plant. These two studies, as well as others, have been instrumental in suggesting that control of urban point source pollution would not substantially eliminate urban pollution as formerly thought, and that adequate attention has to be directed towards research in nonpoint source pollution.

Stable urban areas do not necessarily contribute high suspended sediment loads to streams. However, when an urban area is undergoing development and surfaces lie bare for periods of time, urban runoff with its higher peaks may accelerate erosion and bring about the discharge of high suspended sediment loads to receiving watercourses. Construction of new buildings and roads are two prime contributors to high suspended sediment concentration.

Yorke and Herb (1978) demonstrate the increase in annual sediment yields following construction. Urban land produced about 3.7 tons/ac and this was mainly a result of stream channel erosion downstream of a recently added residential or commercial area (as the authors suggest). Urban construction sites produced an annual yield of 33 tons/ac, which is on the order of ten times that of the urban areas they monitored.

The Vice et al. (1969) study shows that sediment yields following suburban highway construction can be as dramatic as those experienced in urban development (an average annual yield of 63 tons/ac).

These latter two studies show that storms which occupy a small amount of time produce the large majority of total sediment movement.

The concentration of chloride in the urban environment has generated much interest of late, since underground as well as surface waters have been becoming more saline. This occurs because many cities employ street salting in the winter to keep traffic moving safely. Road salts have been shown to be accumulating in urban areas and remaining well after the winter season.

Wulkowicz and Saleem (1974) showed that in their study area about 62.4% of the salt applied was removed in surface flows within the six-month study period. They estimated that about 10% of the remaining 37% was removed by surface flows in the subsequent six-month period, and the remaining portions were removed by subsurface flow or retained in the basin by lag mechanisms.

Because of the lead in gasoline and the number of automobiles in use, lead also concentrates in the urban environment. When it reaches a stream, it may disrupt the flora and fauna and, eventually, affect humans.

Oliver et al. (1974) demonstrate that lead concentrates in urban snow and, on being dumped into snow piles outside of the city, may accumulate to dangerously high concentrations. If lead at these concentration levels should reach a stream, it could seriously deteriorate water quality and create an adverse environmental impact which could be felt all the way up the food chain.

It may be concluded from this chapter that natural streams and natural water quality may be seriously influenced by urban runoff. On par with this conclusion, it must be pointed out that the degree and kind of water quality change is unique to each urban region and must be carefully studied within that context.

3.9 Exercises

Again considering an urban area with which you are very familiar:

1. Which water quality parameters do you think have been affected most by the surface runoff due to the land surface modifications peculiar to this urban area?
2. Select several locations along the stream draining the urban area. Draw a pollutograph (see Figure 3.9) for a hypothetical rainstorm for each location. Construct a table such as Table 3.5 for the timing of the water quality parameters in each pollutograph. Explain differences between the pollutographs and the timings of the chemical constituents.
3. Which water quality parameter in the streamflow experiences the greatest change following a rainstorm? Why? Would the season of year affect the concentrations of the water quality parameters in the stream? Which parameters and how would they be affected? Is there an interaction between the chemical constituents?
4. Compare your pollutographs to those in Figures 3.9–3.11. How do they differ and why are there differences?
5. If water quality data are available for the stream (see Chapter 5 for sources of data), actual pollutographs may be drawn. How does this pollutograph differ from the one(s) above? Why does it differ?
6. Compare the urban area to a nearby urban area. How do the pollutographs differ? Why do they differ?
7. If you were to take samples of the fish and invertebrates at selected points on the stream, what would you expect to find? What do you think the BODs would be at these points?
8. If you were to sample the water quality of the stream during high water flow immediately after a rainstorm, do you think you would find the water quality to be better than the water quality in the stream during low flow before the storm? Why?

References

AMERICAN PUBLIC WORKS ASSOCIATION. "Water Pollution Aspects of Urban Runoff," U.S. Dept. of the Interior, Washington, DC (January 1969).

AMY, G. R. et al. "Water Quality Management Planning for Urban Runoff," U.S. Environmental Protection Agency, EPA #440/9-75-004, Washington, DC (1974).

ARMSTRONG, N. E. et al. "Ecological Aspects of Stream Pollution," in *Advances in Water Quality Improvement*. E. F. Gloyna and W. W. Eckenfelder, Jr., eds. Austin: University of Texas Press, pp. 83–95 (1968).

BARTSCH, A. F. "Biological Aspects of Stream Pollution," in *Biology of Water Pollution. A Collection of Selected Papers on Stream Pollution, Waste Water and Water Treatment*. By L. E. Keup et al. Federal Water Pollution Control Administration, Cincinnati, Ohio, 1948, pp. 13–20 (October 1967).

BARTSCH, A. F. and W. M. Ingram. "Stream Life and the Pollution Environment," *Public Works*, 90:104–110 (1959).

BENOIT, R. J. "Self Purification of Natural Waters," in *Water and Water Pollution Handbook, Vol. 1*. L. L. Ciaccio, ed. New York:Marcel Dekker, Inc., Chapter 4 (1971).

BEST, G. A. and S. L. Ross. *River Pollution Studies*. Liverpool, England:Liverpool University Press, 92 pp. (1977).

BIESECKER, J. E. and D. K. Leifeste. "Water Quality of Hydrologic Bench Marks—As Indicator of Water Quality in the Natural Environment," U.S. Geological Survey Circular #460-E, 21 pp. (1975).

BREZONIK, P. L. "Nitrogen: Sources and Transformations in Natural Waters," in *Nutrients in Natural Waters*. E. Allen and R. Kramer, eds. New York:Wiley-Interscience, pp. 1–50 (1972).

BROWNLEE, R. C. et al. "Variation of Urban Runoff Quality with Duration and Intensity of Storms," Water Resources Center, Texas Tech. University, Lubbock, Texas (1970).

BRYAN, E. H. "Quality of Stormwater Drainage from Urban Land," *Water Resources Bull.*, 8(3):578–588 (1972).

CHERKAUER, D. S. "Urbanization's Impact on Water Quality During a Flood in Small Watersheds," *Water Resources Bull.*, 11(5):987–998 (1975).

COLBY, B. R. "Fluvial Sediments: A Summary of Source, Transportation, Deposition, and Measurement of Sediment Discharge, U.S. Geological Survey Bulletin #1181-A, 47 pp. (1963).

COURCHAINE, R. J. "Significance of Nitrification in Stream Analysis—Effects on the Oxygen Balance," *J. Water Poll. Control Fed.*, 40(5)I:835–847 (1968).

DAWDY, D. R. "Knowledge of Sedimentation in Urban Environments," *J. Hyd. Div., ASCE*, 93 (HY6):235–245 (1967).

DIGIANO, F. A. et al. "A Projection of Pollutional Effects of Urban Runoff in the Green River, Massachusetts," in *Urbanization and Water Quality Control*. W. Whipple, Jr., ed. Minneapolis:American Water Resources Association, pp. 28–37 (1975).

DURUM, W. H. "Chemical, Physical and Biological Characteristics of Water Resources," in *Water and Water Pollution Handbook, Vol. 1*. L. L. Ciaccio, ed. New York:Marcel Dekker, Inc., Chapter 1 (1971).

EATON, J. W. "The Biology of Polluted Waters," in *River Pollution Studies*. G. A. Best and S. L. Ross, eds. Liverpool, England:Liverpool University Press, Chapter 4 (1977).

Federal Water Quality Administration, Department of the Interior. "Storm Water Pollution from Urban Land Activity," Water Pollution Control Research Series, 11034 FKL 07/70, Washington, DC (1970).

FIELD, R. and J. A. Lager. "Urban Runoff Pollution Control—State-of-the-Art," *J. Env. Eng. Div., ASCE*, EE1:107–125 (1975).

GAMBELL, A. W. and D. W. Fisher. "Chemical Composition of Rainfall, Eastern North Carolina and Southeastern Virginia," U.S. Geological Survey Water Supply Paper 1535K, 41 pp. (1966).

GRAF, W. L. "The Impact of Suburbanization on Fluvial Geomorphology," *Water Resources Res.*, 11(5):690–692 (1975).

GUY, H. P. "Residential Construction and Sedimentation at Kensington, Maryland," in *Proceedings of the Federal Inter-Agency Sedimentation Conference*, Agricultural Research Service Miscellaneous Publications No. 970 (1963).

HACK, J. T. "Studies of Longitudinal Stream Profiles in Virginia and Maryland," U.S. Geological Survey Professional Paper #294-B, 97 pp. (1957).

HAMMER, T. R. "Stream Channel Enlargement Due to Urbanization," *Water Resources Res.*, 8 (6):1530–1540 (1972).

HAMMER, T. R. "The Effect of Urbanization on Stream Channel Enlargement," University of Pennsylvania at Philadelphia, #72-6161 (4708-B), 330 pp. (1971).

HAWKINSON, R. O. et al. "Quality of Rivers of the United States, 1974 Water Year—Based on the National Stream Quality Accounting Network (NASQAN)," U.S. Geological Survey Open File Report #77-151, 158 pp. (1977).

HEM, J. D. "Study and Interpretation of the Chemical Characteristics of Natural Water," U.S. Geological Survey Water Supply Paper #1473, 363 pp. (1970).

HYNES, H. B. N. *The Biology of Polluted Waters*. Liverpool, England:Liverpool University Press, 202 pp. (1960).

KELLOGG, W. W. et al. "The Sulfur Cycle," *Science*, (4022):587–595 (1972).

KERR, P. C. et al. "The Carbon Cycle in Aquatic Ecosystems," in *Nutrients in Natural Wastes*. H. E. Allen and J. R. Kramer, eds. New York:Wiley-Interscience, pp. 101–124 (1972).

LANGBEIN, W. B. and W. H. Durum. "The Aeration Capacity of Streams," U.S. Geological Survey Circular #542, 6 pp. (1967).

LANGBEIN, W. B. and L. B. Leopold. "Quasi-Equilibrium States in Channel Morphology," *Am. J. Sci.*, 262:782–794 (1964).

LAXEN, D. P. H. and R. Harrison. "The Highway as a Source of Water Pollution: An Appraisal with the Heavy Metal Lead," *Water Research*, 11:1–11 (1977).

LOEHR, R. C. "Characteristics and Comparative Magnitude of Non-Point Sources," *J. Water Poll. Control Fed.*, 46(8):1849–1871 (1974).

MORISAWA, M. *Streams: Their Dynamics and Morphology*. New York:McGraw-Hill Book Company, 175 pp. (1968).

NOVOTNY, V. et al. "Mathematical Modeling of Land Runoff Contaminated by Phosphorus," *J. Water Poll. Control Fed.*, 50(1):101–112 (1978).

OLIVER, B. G. et al. "Chloride and Lead in Urban Snow," *J. Water Poll. Control Fed.*, 46(4): 766–771 (1974).

OVERTON, D. E. and M. E. Meadows. *Stormwater Modeling*. New York:Academic Press Inc., 358 pp. (1976).

PAINTER, H. A. "Chemical, Physical and Biological Characteristics of Wastes and Waste Effluents," in *Water and Water Pollution Handbook, Vol. 1*. L. L. Ciaccio, ed. New York:Marcel Dekker, Inc., Chapter 7 (1971).

RADZIUL, J. V. et al. "Effects of Non-Point Discharge on Urban Stream Quality," in *Urbanization and Water Quality Control*. W. Whipple, Jr., ed. Minneapolis:American Water Resources Association, pp. 201–210 (1975).

RIMER, A. E. et al. "Characterization and Impact of Stormwater Runoff from Various Land Cover Types," *J. Water Poll. Control Fed.*, 50(2):252–264 (1978).

RUANE, R. J. and P. A. Krenkel. "Nitrification and Other Factors Affecting Nitrogen in the Holston River," *J. Water Poll. Control Fed.*, 50(8):2016–2028 (1978).

SARTOR, J. D. et al. "Water Pollution Aspects of Street Surface Contaminants," *J. Water Poll. Control Fed.*, 46:458–467 (1974).

SINGH, U. P. et al. "Approximate BOD Treatment Requirements for Urban Runoff," *Journal of Environmental Engineering Division*, American Society of Civil Engineers, 105(EE1):3–12 (February, 1979).

SPIEKER, A. M. "Water in Urban Planning," U.S. Geological Survey Water Supply Paper #2002, 147 pp. (1970).

STOKER, H. S. and S. L. Seager. *Environmental Chemistry: Air and Water Pollution*. Glenview, IL:Scott, Foresman and Co., 186 pp. (1972).

STREETER, H. W. and E. B. Phelps. "A Study of the Pollution and Natural Purification of the Ohio River," U.S. Public Health Service, Public Health Bulletin #146, 75 pp. (1925).

SWERDON, P. M. and R. R. Kountz. "Sediment Runoff Control at Highway Construction Sites," Engineering Research Bulletin #B-108, The Pennsylvania State University, College of Engineering, University Park, PA, 70 pp. (1973).

TURK, et al. *Environmental Science*. Philadelphia:W. B. Saunders Co., 563 pp. (1974).

UNITED STATES FEDERAL WATER POLLUTION CONTROL ADMINISTRATION. "Report to the Committee on Water Quality Criteria," U.S. Government Printing Office, 234 pp. (1968).

UNITED STATES PUBLIC HEALTH SERVICES. "Drinking Water Standards," U.S. Public Health Service Publication #956, 61 pp. (1962).

USINGER, R. L. *The Life of Rivers and Streams*. New York:McGraw-Hill Book Company, 232 pp. (1967).

VELZ, C. J. *Applied Stream Sanitation*. New York:John Wiley and Sons, Inc., 619 pp. (1970).

VICE, R. B. et al. "Sediment Movement in an Area of Suburban Highway Construction, Scott Run Basin, Fairfax County, VA, 1961–1964," U.S. Geological Survey Water Supply Paper #1591-E, 41 pp. (1969).

WEIBEL, S. R. et al. "Urban Land Runoff as a Factor in Stream Pollution," *J. Water Poll. Control Fed.*, 36:914–924 (1964).

WEIBEL, S. R. et al. "Pesticides and Other Contaminants in Rainfall and Runoff," *J. Am. Water Works Assoc.*, pp. 1075–1084 (1966).

WHIPPLE, W. J. et al. "Unrecorded Pollution from Urban Runoff," *J. Water Poll. Control Fed.*, 46 (5):873–885 (1974).

WHIPPLE, W. J. et al. "Characterization of Urban Runoff," *Water Resources Res.*, 14(2):370–372 (1978).

WOLMAN, M. G. "A Cycle of Sedimentation and Erosion in Urban River Channels," *Geog. Annaler*, 49A:385–395 (1967).

WOLMAN, M. G. and A. P. Schick. "Effects of Construction on Fluvial Sediment, Urban and Suburban Areas of Maryland," *Water Resorces Res.*, 3:451–464 (1967).

WULKOWICZ, G. M. and Z. A. Saleem. "Chloride Balance of an Urban Basin in the Chicago Area," *Water Resources Res.*, 10(5):974–982 (1974).

YORKE, T. H. and W. J. Herb. "Effects of Urbanization on Streamflow and Sediment Transport in the Rock Creek and Anacostia River Basins, Montgomery County, Maryland, 1962–1974," U.S. Geological Survey Professional Paper #1003, 71 pp. (1978).

ZANONI, A. E. "Waste Water Deoxygenation at Different Temperatures," *Water Resources Res.*, 1: 543–566 (1967).

Selected Readings

Alexander, M. "Environmental Impact of Nitrate, Nitrite, and Nitrosamines," in *The Aquatic Environment*. Symposium sponsored by the U.S. EPA Office of Water Program Operations, Washington, DC, pp. 135–148 (1972).

Austin, J. H. and R. S. Engelbrecht. "Ecological Changes in a Polluted Environment," in *Advances in Water Quality Improvement*. E. F. Gloyna and W. W. Eckenfelder, Jr., eds. Austin: University of Texas Press, pp. 9–20 (1968).

Bansal, M. K. "Nitrification in Natural Streams," *J. Water Poll. Control Fed.*, 48(10):2380–2393 (1976).

Bansal, M. K. "Atmospheric Reaeration in Natural Streams," *Water Res.*, 7:769–782 (1973).

Barkdoll, M. P. et al. "Some Effects of Dustfall on Urban Stormwater Quality," *J. Water Poll. Control Fed.*, 49(9):1976–1984 (1977).

Bedient, P. B. et al. "Low Flow and Stormwater Quality in Urban Channels," *Journal of Environmental Engineering Division*, American Society of Civil Engineers, 106(EE2):421–436 (1980).

Bedient, P. B. et al. "Stormwater Pollutant Load-Runoff Relationships," *Journal of Water Pollution Control Federation*, 52(9):2396–2404 (1980).

Bennett, J. P. and R. E. Rathburn. "Reaeration in Open-Channel Flow," U.S. Geological Survey, Professional Paper 737, 75 pp. (1972).

Bott, T. L. "Bacteria and the Assessment of Water Quality," in *Biological Methods for the Assessment of Water Quality*. Philadelphia: American Society for Testing and Materials, pp. 61–75 (1960).

Bott, T. L. "Nutrient Cycles in Natural Systems: Microbial Involvement," in *Biological Control of Water Pollution*. J. Tourbier and R. W. Pierson, Jr., eds. Pittsburgh: University of Pennsylvania Press, Chapter 7 (1976).

Brunner, F. A. and K. B. Schnelle, Jr. "Air Pollution Patterns in an Urban Street Canyon," *J. Env. Eng. Div.*, ASCE, 100(EE2):311–323 (1974).

Bryan, E. H. "Concentrations of Lead in Urban Stormwater," *J. Water Poll. Control Fed.*, 46(10): 2419–2421 (1974).

Burton, T. M. et al. "The Impact of Highway Construction on a North Florida Watershed," *Water Resources Bull.*, 12(3):529–538 (1976).

Cairns, J., Jr. "Factors Affecting the Number of Species in Fresh-Water Protozoan Communities," in *The Structure and Function of Fresh-Water Microbial Communities*. J. Cairns, Jr., ed. Research Division Monograph No. 3, Virginia Polytechnic Institute, American Microscopical Society Symposium, pp. 219–248 (1969).

Carson, M. A. *The Mechanics of Erosion*. London: Pion Limited, 174 pp. (1971).

Chen, C. L. "Urban Storm Runoff Inlet Hydrology Study, Vol. 1: Computer Analysis of Runoff from Urban Highway Watersheds Under Time- and Space-Varying Rainstorms," Federal Highway Division, Report No. FHWA-RD-76-116 (1976).

Chen, C.-N. "Effect of Land Development on Soil Erosion and Sediment Concentration in an Urbanizing Basin: Effects of Man on the Interface of the Hydrological Cycle with the Physical Environment," IAHS Publication No. 113, pp. 150–157 (1974).

Cherkauer, D. S. "The Hydrologic Response of Small Watersheds to Suburban Development: Observations and Modeling," in *Urbanization and Water Quality Control*. W. Whipple, Jr., ed. Minneapolis: American Water Resources Association, pp. 110–119 (1975).

Cherkauer, D. S. "Effects of Urban Lakes on Quantity and Quality of Baseflow," *Water Resources Res.*, 13(6):1119–1130 (1977).

Cherkauer, D. S. "Effects of Urban Lakes on Surface Runoff and Water Quality," *Water Resources Bull.*, 13(5):1057–1067 (1977).

Cherkauer, D. S. and N. A. Ostenso. "The Effect of Salt on Small, Artificial Lakes," *Water Resources Bull.*, 12(6):1259–1266 (1976).

Churchill, M. A. et al. "The Prediction of Stream Aeration Rates," *J. San. Eng. Div.*, ASCE, 88 (SA4):1–46 (1962).

Collins, P. G. and J. W. Ridgway. "Urban Storm Runoff Quality in Southeast Michigan," *Journal of Environmental Engineering Division*, American Society of Civil Engineers, 106(EE1): 153–162 (1980).

Colston, N. V. and A. N. Tafuri. "Urban Land Runoff Considerations," in *Urbanization and Water Quality Control*. W. Whipple, Jr., ed. Minneapolis: American Water Resources Association, pp. 120–128 (1975).

Cordery, I. "Quality Characteristics of Urban Stormwater in Sydney, Australia," *Water Resources Res.*, 13(1):197–202 (1977).

Cowen, W. F. and G. F. Lee. "Leaves as Sources of Phosphorous," *Environ. Sci. Technol.*, 7(9): 853–854 (1973).

Cowen, W. F. et al. "Nitrogen Availability in Urban Runoff," *J. Water Poll. Control Fed.*, 48(2): 339–345 (1976).

Cowen, W. F. "Phosphorus Availability in Particulate Materials Transported by Urban Runoff," *J. Water Poll. Control Fed.*, 48(3):580–591 (1976).

Crippen, J. R. "Change in Quantity of Dissolved Solids Transported by Sharon Creek Near Palo Alto, California, After Suburban Development," U.S. Geological Survey Professional Paper 575-D, pp. D256–D258 (1967).

Dugan, P. *Biochemical Ecology of Pollution*. New York: Plenum Publishing Corporation, 159 pp. (1972).

Dyhouse, G. R. "Sediment Analyses for Urbanizing Watersheds," *Journal of Hydraulics Division*, American Society of Civil Engineers, 108(HY3):399–418 (1982).

Eckenfelder, W. W. and D. L. Ford. "New Concepts in Oxygen Transfer and Aeration," in *Advances in Water Quality Improvement*. E. F. Gloyna and W. W. Eckenfelder, Jr., eds. Austin: University of Texas Press, pp. 215–236 (1968).

Ellis, M. M. "Detection and Measurements of Stream Pollution," in *Biology of Water Pollution*. FWPCA, U.S. Department of the Interior, pp. 129–186 (1967).

Environmental Science. "Urban Runoff Adds to Water Pollution," Editorial Comment. *Environ. Sci.*, 3(6):527 (1969).

Envirogenics Company. "Urban Storm Runoff and Combined Sewer Overflow Pollution," Environmental Protection Agency Water Pollution Control Resource Service, No. 11024, FKM 12/71 (1971).

Feuerstein, D. L. and A. O. Freidland. "Pollution in San Francisco's Urban Runoff," in *Urbanization and Water Quality Control*. W. Whipple, Jr., ed. Minneapolis: American Water Resources Association, pp. 85–94 (1975).

Field, R. and R. Turkeltaub. "Urban Runoff Receiving Water Impacts Program Overview," *J. of Environmental Engineering Division*, American Society of Civil Engineers, 107(EE1):83–100 (1981).

Franke, O. L. and N. E. McClymonds. "Summary of the Hydraulic Situation on Long Island, New York, as a Guide to Water-Management Alternatives," U.S. Geological Survey Professional Paper 627F, Washington, DC, 59 pp. (1972).

Frankel, R. J. and W. W. Hansen. "Biological and Physical Responses in a Freshwater Dissolved Oxygen Model," in *Advances in Water Quality Improvement*. E. F. Gloyna and W. W. Eckenfelder, Jr., eds. Austin: University of Texas Press, pp. 126–142 (1968).

Fruh, E. G. "Urban Effects on Quality of Streamflow," in *Water Resources Symposium Number Two*, Center for Research in Water Resources. Austin: The University of Texas, pp. 255–281 (1969).

Gates, C. D. "Urban Runoff in Binghamton, New York," in *Urbanization and Water Quality Control*. W. Whipple, Jr., ed. Minneapolis:American Water Resources , pp. 38–44 (1975).

Geldreich, E. E. and B. A. Kenner. "Concepts of Fecal Streptococci in Stream Pollution," *J. Water Poll. Control Fed.*, 41(8)2:R336–R352 (1969).

Glymph, L. M. "Importance of Sheet Erosion as a Source of Sediment," *Trans. Am. Geophys. Union*, pp. 903–907 (1956).

Grava, S. *Urban Planning Aspects of Water Pollution Control*. New York:Columbia University Press, 223 pp. (1969).

Guarraia, L. J. "Interrelationship of the Major Nutrients and Microorganisms, in *The Aquatic Environment*. Symposium sponsored by the EPA Office of Water Program Operations, Washington, DC, pp. 3–8 (1972).

Gundelach, J. M. and J. E. Castillo. "Natural Stream Purification under Anaerobic Conditions," *J. Water Poll. Control Fed.*, 48(7):1753–1758 (1976).

Hawkins, R. H. and J. H. Judd. "Water Pollution as Affected by Street Salting," *Water Resources Bull.*, 8(6):1246–1252 (1972).

Heaney, J. P. et al. "Nationwide Evaluation of Combined Sewer Overflows and Urban Stormwater Discharges," EPA-600/2-77-064, Environmental Protection Technology Series, Environmental Protection Agency (1977).

Hynes, H. B. N. *The Ecology of Running Waters*. Toronto:University of Toronto Press, 555 pp. (1970).

Ignjatovic, L. R. "Effect of Photosynthesis on Oxygen Saturation," *J. Water Poll. Control Fed.*, 40(5)2:R151–161 (1968).

Johnson, R. E. et al. "Dustfall as a Source of Water Quality Impairment," *J. Env. Eng. Div.*, ASCE, 92(SA1):245–267 (1966).

Katz, M. "The Effects of Pollution Upon Aquatic Life," in *Water and Water Pollution Handbook*, Vol. 1. L. L. Ciaccio, ed. New York:Marcel Dekker, Inc., Chapter 6 (1971).

Keller, F. J. "Effect of Urban Growth on Sediment Discharge in the Northwest Branch of the Anacostia River Basin," U.S. Geological Survey, Professional Paper 450-C, pp. C129–C131 (1962).

Kholdebarin, B. and J. J. Oertli. "Effect of Suspended Particles and Their Sizes on Nitrification in Surface Water," *J. Water Poll. Control Fed.*, 49(7):1693–1697 (1977).

Kholdebarin, B. "Effect of pH and Ammonia on the Rate of Nitrification of Surface Water," *J. Water Poll. Control Fed.*, 49(7):1688–1692 (1977).

Koch, E. "The Effects of Urbanization on the Quality of Selected Streams in Southeastern Nassau County, Long Island," U.S. Geological Survey, Professional Paper 700C, pp. 189192 (1970).

Komura, S. "Hydraulics of Slope Erosion by Overland Flow," *J. Hyd. Div.*, ASCE, (HY10): 1573–1586 (1976).

Kramer, J. R. et al. "Phosphorus: Analysis of Water, Biomass and Sediment," in *Nutrients in Natural Waters*. H. E. Allen and J. R. Kramer, eds. New York:Wiley Interscience, pp. 51–100 (1972).

LeClerc, E. "Self-Purification of Fresh Water Streams as Affected by Temperature and by the Content of Oxygen, Nitrogen and Other Substances," *Int. J. Air Water Poll.*, 7:357–365 (1963).

Legall, J. "The Sulfur Cycle," in *The Aquatic Environment*. Symposium sponsored by the Environmental Protection Agency Office of Water Programs Operations, Washington, DC, pp. 75–86 (1972).

Linzon, S. N. "Pre-Pollution Background Studies in Ontario," *J. Soil Water Cons.*, 28(5):226–228 (1973).

Livingstone, D. A. "Chemical Composition of Rivers and Lakes," in *Data of Geochemistry*. 6th edition. M. Fleischer, ed. U.S. Geological Survey Professional Paper 440-G (1963).

MacKenthun, K. M. "The Practice of Water Pollution Biology," U.S. Department of the Interior, Federal Water Pollution Control Administration, Division of Technical Support (1969).

Maguire, B., Jr. "Community Structure of Protozoans and Algae with Particular Emphasis on Recently Colonized Bodies of Water," in *The Structure and Function of Fresh-Water Microbial Communities*. J. Cairns, Jr., ed. Research Division, Monograph 3, Virginia Polytechnic Institute, American Microscopical Society Symposium, pp. 121–150 (1969).

Mattraw, H. C., Jr. and C. B. Sherwood. "Quality of Stormwater Runoff from a Residential Area, Broward County, Florida," *J. Res. U.S. Geological Survey*, 5(6):823–834 (1977).

Menard, H. W. "Some Rates of Regional Erosion," *J. Geol.*, 69(2):154–161 (1961).

Middleton, A. C. and A. W. Lawrence. "Kinetics of Microbial Sulfate Reduction," *J. Water Poll. Control Fed.*, 49(7):1659–1670 (1977).

Miller, R. A. and H. C. Mattraw, Jr. "Storm Water Runoff Quality from Three Land-Use Areas in South Florida," *Water Resources Bulletin*, 18(3):513–519 (June, 1982).

Mitchell, R. *Water Pollution Microbiology*. New York:Wiley Interscience, 416 pp. (1972).

Moffa, P. E. et al. "Urban Runoff and Combined Sewer Overflow," *Journal of the Water Pollution Control Federation*, 54(6):675–677 (1982).

O'Connor, D. J. and D. M. DiToro. "An Analysis of the Dissolved Oxygen Variation in a Flowing Stream," in *Advances in Water Quality Improvement*. E. F. Gloyna and W. W. Eckenfelder, Jr., eds. Austin:University of Texas Press, pp. 96–102 (1968).

O'Connor, D. J. and W. E. Dobbins. "Mechanisms of Reaeration in Natural Streams," *ASCE Trans.*, 123:641–684 (1958).

Palmer, C. L. "The Pollutional Effects of Stormwater Overflows from Combined Sewers," presented at the 24th Annual Conference, Michigan Sewage Works Association, Traverse City, Michigan (May 23–25, 1949).

Payne, W. J. "Bacterial Growth Yields," in *The Aquatic Environment*. Symposium sponsored by the Environmental Protection Agency, Office of Water Program Operations, Washington, DC, pp. 57–74 (1972).

Ragan, R. M. and A. J. Dietemann. "Impact of Urban Stormwater Runoff on Stream Quality," in *Urbanization and Water Quality Control*. W. Whipple, Jr., ed. Minneapolis:American Water Resources Association, pp. 55–61 (1975).

Randall, C. W. et al. "Characterization of Urban Runoff in the Occoquan Watershed of Virginia," in *Urbanization and Water Quality Control*. W. Whipple, Jr., ed. Minneapolis:American Water Resources Association, pp. 62–69 (1975).

Reid, G. K. *Ecology of Inland Waters and Estuaries*. New York:Reinhold Publishing Corporation, 375 pp. (1961).

Rickert, D. A. and W. G. Hines. "River Quality Assessment: Implications of a Prototype Project," *Science*, 200(4346):1113–1117 (1978).

Saunders, G. W. "Carbon Flow in the Aquatic System," in *The Structure and Function of Fresh-Water Microbial Communities*. J. Cairns, Jr., ed. Research Division Monograph #3, Virginia Polytechnic Institute, American Microscopical Society Symposium, pp. 31–46 (1969).

Schumm, S. A. *The Fluvial System*. New York:Wiley-Interscience Publications, 338 pp. (1977).

Sedimentation Committee of the Water Resource Council. *Proceedings of the Third Federal*

Interagency Sedimentation Conference, 1976, Denver, Colorado, March 22-25. No. PB245-100, 974 pp. (1975).

Sokatch, J. R. *Bacterial Physiology and Metabolism.* London:Academic Press, Inc., 443 pp. (1969).

Stack, V. T., Jr. "Stabilization Oxygen Demand," in *Biological Methods for the Assessment of Water Quality.* Philadelphia:American Society for Testing and Materials, pp. 209-220 (1969).

Stenstrom, M. K. et al. "Oil and Grease in Urban Storm Waters," *Journal of Environmental Engineering Division,* American Society of Civil Engineering, 110(EE1):58-72 (1984).

Sylvester, M. A. and W. M. Brown, III. "Relation of Urban Landuse and Land-Surface Characteristics to Quantity and Quality of Storm Runoff in Two Basins in California," U.S. Geological Survey, Water Supply Paper 2051, 49 pp. (1978).

Tuffey, T. J. et al. "Zones of Nitrification," *Water Resources Res.,* 10(3):555-564 (1974).

Umbreit, W. W. and the Class of '75. *Essentials of Bacterial Physiology.* Stroudsburg, PA:Dowden Hutchinson and Ross, Inc. (1976).

Vice, R. B. et al. "Erosion from Suburban Highway Construction," *ASCE Proc.,* Paper #5742, 94: 347-348 (1968).

Walling, D. E. and K. J. Gregory. "The Measurement of the Effects of Building Construction on Drainage Basin Dynamics," *J. Hydrology,* 11:129-144 (1970).

Wezernak, C. T. and J. J. Gannon. "Evaluation of Nitrification in Streams," ASCE, *J. San. Eng. Div.,* 94(SA5):883-895 (1968).

Whipple, W., Jr. "BOD Mass Balance and Water Quality Standards," *Water Resources Res.,* 6(3): 827-837 (1970).

Whipple, W., Jr., ed. *Urban Runoff—Quantity and Quality.* New York:Am. Soc. Civil Engineers, 272 pp. (1975a).

Whipple, W., Jr., ed. *Urbanization and Water Quality Control.* Minneapolis:American Water Resources Association, 302 pp. (1975b).

Whipple, W., Jr. "Strategy of Water Pollution Control," *Water Resources Bull.,* 12(5):1019-1028 (1976).

Whipple, W., Jr. et al. "Effects of Storm Frequency on Pollution from Urban Runoff," *J. Water Poll. Control Fed.,* 49(11):2243-2248 (1977).

Whipple, W., Jr. et al. "Runoff Pollution from Multiple Family Housing," *Water Resources Res.,* 14(2):288-301 (1978).

Whipple, W., Jr., J. M. DiLouie and T. Pytlar, Jr. "Erosional Potential of Streams in Urbanizing Areas," *Water Resources Bulletin,* 17(1):36-45 (February 1981).

Wilber, W. G. and J. V. Hunter. "Contributions of Metals Resulting from Stormwater Runoff and Precipitation in Lodi, New Jersey," in *Urbanization and Water Quality Control.* W. Whipple, Jr., ed. Minneapolis:American Water Resources Association, pp. 45-54 (1975).

Wilber, W. G. and J. V. Hunter. "Distribution of Metals in Street Sweepings, Stormwater Solids and Urban Aquatic Sediments," *Journal of the Water Pollution Control Federation,* 51(12): 2810-2822 (1979).

Wischmeier, W. H. and J. V. Mannering. "Relation of Soil Properties to its Erodibility," *Soil Sci.,* pp. 131-137 (1969).

Wolman, M. G. "Erosion in the Urban Environment," *Hydrologic Sci. Bull.,* 20(1):117-125 (1975).

Wuhrmann, K. "Stream Purification," in *Water Pollution Microbiology*. R. Mitchell, ed. New York:Wiley Interscience, Chapter 6 (1972).

Yu, S. L. et al. "Assessing Unrecorded Organic Pollution from Agricultural, Urban and Wooded Lands," *Water Res.*, 9:849-852 (1975).

Zogorski, J. S. et al. "Temporal Characteristics of Stormwater Runoff: An Overview," in *Urbanization and Water Quality Control*. W. Whipple, Jr., ed. Minneapolis:American Water Resources Association, pp. 100-109 (1975).

4 Analysis of Hydrologic Change Due to Urbanization

> . . . the distinctive hydrologic characteristic of urban hydrology might be said to be the change in runoff response of an area as a function of its development. Thus, the hydrologic analyst is faced with evaluating the effect of various physical changes in the area.—C. W. Timberman (1970), p. 1.

4.1 How to Start an Urban Hydrologic Study

CHAPTERS 1, 2 and 3 have presented urbanization and its impact on water quantity and water quality, respectively. These chapters have given the reader a brief conceptual understanding of urban hydrological problems, in essence, the background necessary to begin an urban hydrologic study. This chapter will demonstrate various methods which can be employed to define the extent of urban land use change and various graphic means of portraying the changed hydrologic response of the watershed.

This entire book has been written to emphasize the major points in urban hydrology. The chapters all present a brief overview, and attempt within the limited review to be comprehensive. However, a compromise had to be established which allowed the major points to be emphasized in a minimum amount of space. Each chapter tries to utilize the main sources for the particular topics presented. These references may be used and are recommended for use as a springboard to further research. At the end of each chapter, the Selected Readings list numerous unreferenced sources, which the author has perused in his structuring and writing of this book. These references may furnish the reader with additional pertinent urban hydrological research which he or she may find helpful.

In beginning an urban hydrologic study, the researcher must first decide in what field his study lies, i.e., water quantity (streamflow related) or water quality (water pollution related). The two sections which follow (4.2 and 4.3) will aid the reader in gathering data in either of these fields. This section will outline several study methods, and then in the remainder a listing of references will be presented which, in the author's opinion, are basic to the study of urban hydrology.

STUDY APPROACHES IN URBAN HYDROLOGY

Snyder (1971) observed that:

> ... detection of hydrologic change is taken to mean analysis of historical records to establish the plausibility of nonhomogeneity of data. Such nonhomogeneity is conventionally associated with some watershed physical change (p. 311).

Urban hydrologic studies usually investigate the effect of land use change on water quantity or water quality. In order to detect this change, three basic approaches may be utilized: (1) upstream-downstream, (2) before and after, or (3) paired watershed. The upstream-downstream method entails the comparison of data collected from a river upstream of the urban area with those collected downstream (along the same river). This method must take into account the addition of tributaries in between the two data collection points and changes in geology. Employment of this method produces very clear and essentially irrefutable evidence of urban land use changes on the study stream.

In the before/after approach, data collected from an urban area prior to the land use change are compared to data collected after the land use change. Results from this method suffer because the researcher cannot totally separate the effects of rainstorms in one period from those in another period. The author (Lazaro, 1976) has applied nonparametric statistics in an attempt to eliminate the effects of precipitation (Section 4.7).

The paired watershed method entails the comparison of hydrologic data from two or more watersheds; one is the urban and the other rural (a control). Data are from concurrent time periods, and any change in these data is taken as being indicative of land use change. The paired watershed method, if properly applied, produces a sound argument. However, the basins compared must be in proximity (for climatic purposes), the geology should be the same, and land use must be stable over the study period in the rural basin.

DEFINITION OF URBANIZING PERIOD

The time period of the change in land use may be determined by observing data on population shifts, or residential or other construction. These data can be found at local planning commissions, at college libraries, and in some cases, at county or city libraries. Population and housing data are collected and enumerated decenially by the U.S. Department of Commerce, Bureau of the Census. Local planning commissions monitor land use changes with great care, and have professional planners on their staffs through whom much information on land use changes can be procured.

SCALE OF THE STUDY AREA

The scale of the study area depends on such factors as: (1) data needs and data availability, (2) study method(s) employed, (3) focus of the study, and (4) time constraints of the study. A researcher has to optimize the scope to conduct a study which will yield sound results, and yet justify expenses and methods of approach. This optimization procedure holds true for almost any study. The purpose of the study must be clearly defined, and then the researcher must select a workable scale. If the area is large, the effects of climate and geology must be properly addressed. Some urban areas like Los Angeles and New York City may fall into the latter category.

BACKGROUND REFERENCES FOR URBAN HYDROLOGY

Most universities have civil engineering, environmental science, geography or geology libraries, and the following texts are usually found in one or all of these libraries.

- *The Handbook of Applied Hydrology*, edited by Ven Te Chow, consists of twenty-nine comprehensive papers which describe the hydrologic cycle, present methods of analysis, data collection techniques, and are generally sound engineering bases for the science of hydrology.
- *Water, Earth and Man*, edited by R. J. Chorley, is a series of essays on the hydrologic cycle from a geographic perspective. It is a fine treatment from a man-oriented approach.
- *Drainage Basin Form and Process*, by K. J. Gregory and D. E. Walling, discusses the watershed from a geomorphological perspective. There is a forty-page bibliography, and also a large amount of information on the physical processes occurring within a catchment.
- *Applied Hydrology*, by R. K. Linsley, Jr., M. A. Kohler and J. L. H Paulus, although an old text (1949), is still held in reverence by present-day hydrologists and referenced quite frequently. It presents a fine treatment of the basic hydrologic concepts.
- *Introduction to Hydrology*, by W. Viessman, Jr. et al., is an up-to-date and comprehensive study of pertinent hydrologic topics. It is one of the better, if not the best, basic hydrologic texts available today.
- *Prediction in Catchment Hydrology*, edited by T. G. Chapman and F. X. Dunin, consists of a compilation of papers from a symposium on hydrology, containing much valuable information.
- *Urbanization and Environment*, edited by T. R. Detwyler and M. G. Marcus, presents many topics which are not hydrologically oriented, but the essays provide a sound overall view of the environmental problems created by urban development.

U.S. GEOLOGICAL SURVEY

The Water Resources Division of the U.S. Geological Survey maintains offices across the United States. At many of these offices, a library of Geological Survey publications may be found. These consist of:

(1) Professional papers, which deal mainly with hydraulic, hydrologic and geologic research
(2) Water supply papers, which also are sometimes research-oriented, but do have numerous papers on hydrologic case studies
(3) Bulletins, which are usually brief presentations of research by noted authorities
(4) Circulars, which contain pertinent discussions on a usually easily understandable level on topics which are deemed to be of national interest

The reader is advised to seek out his or her nearest U.S.G.S. office; not only are numerous hydrologic papers available there, but practicing hydrologists are also available for free consultation. These professionals can guide the reader to information sources and/or provide the researcher with data to conduct a study. A phone call to these scientists may prove to be a worthwhile undertaking, and might save the would-be researcher much effort.

If no U.S.G.S. office is within reasonable access, then the reader may buy many U.S.G.S. publications from U.S. Geological Survey, Box 25425, Federal Center, Denver, CO 80225. Some of the publications are available free of charge. One of the first requests should be for Geological Survey Circular No. 554 – *Hydrology for Urban Land Planning – A Guidebook on Hydrologic Effects of Urban Land Use* by L. B. Leopold. This booklet is a synopsis of urban hydrologic (quantity and quality) problems. It presents many easy-to-interpret graphs and discusses pertinent topics.

4.2 Water Quantity Data Collection

PRIMARY DATA COLLECTION[26]

In Section 2.1 we presented the stream discharge formula as:

$$Q = VA$$

where

Q = discharge in cubic feet per second
V = velocity in feet per second
A = area in square feet

---

[26]Primary data is defined as data collected by the researcher.

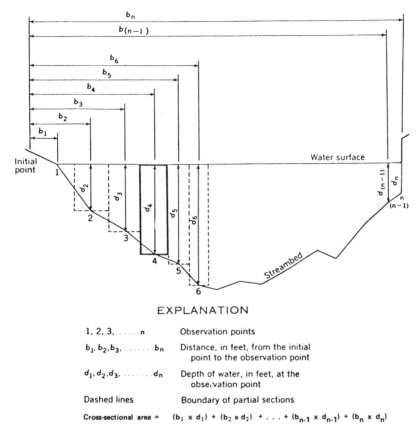

Figure 4.1. Cross section of a stream [*Source*: Buchanan and Somers (1969) p. 2].

Therefore, in order to measure the discharge at a particular cross section along a stream, the average velocity in a cross section of the stream and the area of that cross section must be determined. This is accomplished by dividing the stream into preferably twenty to thirty separate sections, noting each width and measuring the depth and velocity in each (Figure 4.1). By multiplying the width of each section by its depth, we obtain its area. The multiplication of this product by the average velocity yields discharge for that section. The summation of all section discharges equals the stream discharge at that cross section.[27]

As the depth of the river varies at a given cross section, so does discharge for that section. The height of the water surface above an arbitrary (fixed)

---

[27]For a much more detailed description of stream discharge measurements, see Buchanan and Somers (1969).

datum is termed "stage" (which is directly related to river depth; see Buchanan and Somers, 1968). Discharge measuring agencies (e.g., the U.S. Geological Survey, the U.S. Army Corps of Engineers and other federal and state agencies) make periodic streamflow measurements at established stations and plot these measurements on a graph. The ordinate (y-axis) of the graph is stage in feet, and the abscissa is in discharge increments. Such a plot of discharge measurements is used to define the stage/discharge relationship for the particular station, and a rating table is developed from the curve.

Discharge measuring agencies maintain a network of streamflow stations. Most stations have an enclosure which contains continuous-stage monitoring equipment. Along any given river, there are points where accurate discharge measurements may not be obtained. This is due to various hydrologic and hydraulic characteristics of the stream. Therefore, the location of a streamflow monitoring station has to be carefully considered before the station is established and constructed. It is advisable before the reader attempts to measure discharge that he or she consult with a streamflow monitoring agency to define points along a river where streamflow may be measured confidently and accurately.

A simple method of determining streamflow, although a crude one, and not as elaborate as methods utilized by professional agencies, is to use a floating object such as a cork, place it in the stream, and measure the distance in feet traveled in the time elapsed. In this manner, one can derive an approximation of flow velocity. The area of the stream can be estimated by the section method described above, or a modification of it, depending on the accuracy desired.

It must be cautioned that streamflow measurements can be dangerous. Professional streamflow measuring agencies have adopted procedures aimed at preventing injury and lowering the risks to their personnel. In addition, hydrologists measure streamflow at specified locations where a history of the stream channel has been established. The reader is cautioned and advised to seek advice on any stream before attempting to gather data. Urban streams may be "flashy," creating a serious flood hazard. In addition, they may be grossly polluted, which will render them unsuitable for data collection which requires wading and/or any body contact with the water.

SECONDARY DATA COLLECTION[28]

From the network of streamflow measuring agencies comes a continuous supply of discharge data. Stations are closed, reopened, and new ones are constructed and some are modified. The researcher must find a station which monitors streamflow from the study area, and then find out from the discharge

---

[28]Secondary data are published data. Primary data are data (i.e., streamflow measurements, water quality samples, etc.) collected by the researcher.

monitoring agency what the period of record (period of operation) is for the station. The majority of stations may provide a continuous digital tape with a stage punch every fifteen minutes. Through analysis of these tapes, the discharge monitoring agency determines mean daily discharge and, if a storm occurs, a very accurate hydrograph may be plotted. Some stations are equipped with strip chart recorders which continuously plot the stage hydrograph.

The U.S. Geological Survey publishes the mean daily discharge values for its network. These reports usually concern one state, and are free to the public as a service which this agency provides. The annual report concerns streamflow during a water year (which is now concurrent with a fiscal year), which is the period from October 1 to September 30. Streamflow data on a more frequent basis is available from the U.S.G.S. upon request.

Precipitation and other weather data may also be obtained at many U.S.G.S. offices. These data are collected and compiled by the U.S. Weather Bureau and may also be found in most college libraries.

4.3 Water Quality Data Collection

PRIMARY DATA COLLECTION[29]

We will divide water quality sampling (see Hem, 1970; Rainwater and Thatcher, 1960) into four broad categories: (1) dissolved gases and metals, (2) organics, (3) suspended sediment[30] and (4) bacteriological. There are three basic methods employed to collect water quality samples: (1) ETR (Equal Transit Rate), or EWI (Equal Width Increment), (2) EDI (Equal Discharge Increment),[31] and (3) the least accurate point, grab or dip sample. The method employed is in accordance with the hydraulics of the sampled stream. The objective is to get a representative sample.

The ETR method consists of using a uniform rate of lowering (submerging) and raising the sampler through each vertical (section); once this rate is established, it is maintained during the sampling of the remaining verticals. Each vertical has an equal width, and the number of verticals for a stream varies according to the width of the stream and other factors.

The EDI method consists of dividing the stream cross section into subareas

---

[29]It must be reiterated at this point that collecting data from urban streams can be dangerous because of pollutants, broken glass in the streambed, and other factors. Every precaution should be undertaken to reduce safety hazards.

[30]Suspended sediment is classified under water quality in this section. However, it is not necessarily to be considered a water quality indicator.

[31]See Interagency Committee on Water Resources (1963) for a discussion of these methods.

of equal discharge, and taking about equal sample volumes from each subarea. This method usually consists of sampling about four verticals, which is less time-consuming than the ETR method. However, if the centroids of discharge are not accurately known, a discharge measurement should precede sample collection.

The dip sampling method consists of filling a container held just under the water surface. The bottle is secured to a weighted holder and submerged. In shallow streams where the use of the ETR and EDI methods is not possible, dip sampling may produce accurate results. However, care must be taken in not disturbing the bed of the stream during sampling, because bed material might contaminate the sample.

If there are any questions on methods that should be employed or site selection, the U.S Geological Survey or other professional water quality/water quantity monitoring agencies may give assistance.

Dissolved Gases and Metals

The time-varying nature of dissolved gases, pH, specific conductance, and temperature (these constituents may be termed "nonconservatives") requires their measurement at the sampling site (U.S. Geological Survey, Office of Water Data Coordination, 1977).

The collection of dissolved metals is discussed by Brown et al. (1974). Sample containers should be thoroughly cleansed with phosphate-free soap. In some cases, to eliminate interference by external sources prior to sample collection the bottles are rinsed in the river water at the site. After collection, samples for metal analysis can be preserved by the addition of an acid, and/or chilling in an ice bath.

Organics

Some organic substances discussed by Goerlitz and Brown (1972) are carbon, nitrogen and its derivatives, oxygen demand, chemical oxygen demand (COD), pesticides, and insecticides. Glass bottles are the most acceptable containers for collecting, transporting, and storing these samples. Bottles should be thoroughly cleansed and sterilized following a procedure outlined by the authors. Organic samples are particularly susceptible to degradation and must be packed in ice and/or collection should be timed so that the samples may reach the laboratory in a minimum amount of time. Samples to be analyzed for chemical oxygen demand, organic carbon, oils and waxes, and herbicides, may be preserved by the addition of sulfuric acid (Goerlitz and Brown, 1972). Water samples to be analyzed for organics require special handling, and every effort must be made to prevent contamination.

Suspended Sediment

Guy and Norman (1970) discuss the measurement of fluvial sediment. Suspended sediment consists of silts, clays, sands and some organics found suspended in stream waters. These particulates are mainly composed of inorganic debris, and therefore storage after collection is not a crucial factor. Thoroughly cleaned glass pint bottles are usually employed for collection and storage of these samples. Strict adherence to sampling procedures such as the EWI or the EDI is the most crucial part of collection of suspended sediment samples.

Bacteria

Microbiological sampling is presented by Slack et al. (1973), and Greeson et al. (1977). Ideally, the living habits of the organisms to be sampled and their life cycles should be determined, a sound statistical sampling method developed, and a relatively large number of samples taken. In practice, however, the latter combined with sampling procedures such as the EWI and the EDI previously described cannot be accomplished. This is primarily because it is extremely difficult to maintain sterile conditions while sampling, and under nonsterile conditions the samples must be processed immediately after collection. These two factors preclude the employment of multivertical sampling procedures. Therefore, the dip or point sample is most often employed in bacteriological sampling.[32]

Water samples to be analyzed for bacteria are collected in bottles which have been thoroughly cleaned and sterilized. Following the membrane filter technique which has attained widespread application, the samples are filtered, and the filter is placed in an appropriate nutrient medium. Cultures of the bacteria are incubated in either a solid (agar) or liquid (broth) medium, and examined 24–48 hr later (Greeson et al., 1977).

From the brief outline above, it becomes apparent that skilled manpower is needed to conduct an adequate water quality data collection program. In addition, expensive specialized equipment is required. Taking both these factors into account, a major consideration in initial study design is as Alley (1976) summarizes: ". . . water quality data are difficult and expensive to collect, and involve a number of considerations. Only data that are necessary to satisfy the objectives of the study should be collected" (p. 81).

SECONDARY DATA COLLECTION

At many of the streamflow stations discussed above under Water Quantity,

---

[32]The U.S. Geological Survey Office of Water Quality Data Report (1977) points out that a large number of samples are collected using this technique.

the streamflow monitoring agencies collect and may provide to the public free of charge large amounts of water quality data.

Texts for Water Quality

- *River Pollution Studies*, by G. A. Best and S. L. Ross, is a brief, pertinent, and clear handbook on tests for the various water quality constituents. It has a good discussion concerning the causes of river pollution.
- *A Practical Guide to Water Quality Studies of Streams*, by F. W. Kittrell, is also a brief, but pertinent discussion of methods in water quality studies of streams.
- *Water and Water Pollution Handbook*, edited by L. L. Ciaccio, is a comprehensive, extensive treatment of water pollution.
- *Applied Stream Sanitation*, by C. J. Velz, is an excellent discussion of organic and inorganic water quality constituents. A fine text, referenced quite often in the literature.

4.4 The Urbanized Area[33] and the Watershed

A watershed is a natural physiographic region, bounded by a topographic divide. To arrive at estimates of the population contained within the region, one must use political or census-demarcated areas, which usually are not coincident with the drainage divide. Census tract data are the most accurate means of enumerating population, because tracts usually occupy a rather small area, and several may even be found within small urban watersheds. In an urban hydrologic study involving the Northeast Branch of the Anacostia River in the Washington, DC metropolitan area, the author (Lazaro, 1976a) found that census tract data was not available, and therefore used election district data which predated the implementation of census tracts by the Census Bureau. This provided an appropriate means of displaying and defining the population density changes accompanying urbanization. Election district boundaries were reduced to the same scale, and superimposed (Figure 4.2) on a drainage area map. Table 4.1 shows the population by the ten election districts. Figures 4.2 and 4.3 give a spatial demonstration of population densities and shifts in 1940 and 1970, respectively.

---

[33]The U.S. Bureau of Census (1980) has listed several definitions of an urbanized area: 1) a central area of 50,000 persons or more; 2) incorporated places having at least 2500 persons, and surrounding an area as in 1); and 3) small parcels of land in an area having 1000 or more inhabitants per square mile.

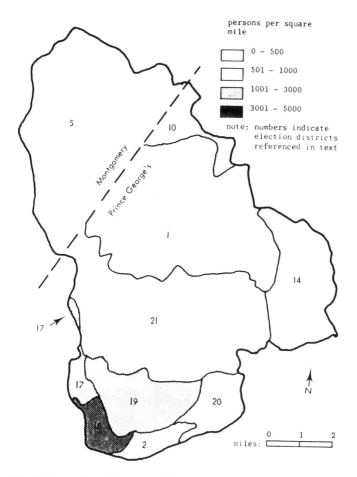

Figure 4.2. Northeast Branch basin population density by election district: 1940 [*Source*: Lazaro, *J. Env. Systems*, "Planning Aids to Measure the Physical Impact of Urbanization," 6(1): 69–82 (1976), p. 71, © 1976, Baywood Publishing Co., Inc.].

In many cases, the majority of land use change accompanying urban development is caused by residential construction. Local planning commissions and/or housing authorities may have housing data; if they do not, the U.S. Bureau of Census enumerates such data. Housing-unit data are usually categorized, i.e., multi-family homes, single-family homes. If a researcher can determine the average area occupied by each type of housing unit, he or she may derive an estimate of the impervious area added by census tract or election district or other suitable unit. Table 4.2 lists housing data for the drainage area discussed above. There was an increase of almost 500% in the number of housing units over the 30-yr study period. A categorical breakdown of the

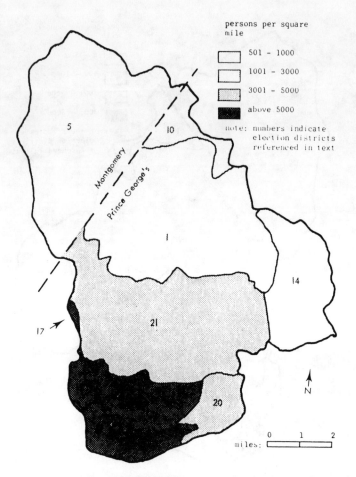

Figure 4.3. Northeast Branch basin population density by election district: 1970 [*Source*: Lazaro, *J. Env. Systems*, "Planning Aids to Measure the Physical Impact of Urbanization," 6(1): 69–82 (1976), p. 73, ©1976, Baywood Publishing Co., Inc.].

types of housing units was available; however, the author used other, more accurate methods to arrive at estimates of land use change.

Aerial photography was employed by the author to derive estimates of land use change for the Northeast Branch Basin. Photography was found that was taken in 1936 and 1968. By carefully examining and listing the various land uses, the author arrived at an accurate estimate of the area occupied by each land use. Harris and Rantz (1964) used similar methods (aerial photography for 1948 and 1960). The land use change in aerial photographs is quite visible when the study area is small enough to be contained in one photograph.

Table 4.1. Comparison of population data for the Northeast Branch Basin 1940 and 1970.

Election District No.	Election District	Population	
		1940	1970
1	Vansville	1,923	20,914
2	Bladensburg	6,103	41,885
10	Laurel	3,691	31,579
14	Bowie	3,600	29,161
16	Hyattsville	7,923	15,491
17	Chillum	10,864	75,728
19	Riverdale	6,187	21,909
20	Lanham	1,758	37,739
21	Berwyn	7,741	61,688
5	Colesville	4,045	49,605
Totals		53,835	385,699

Source: Lazaro, J. Env. Systems, "Planning Aids to Measure the Physical Impact of Urbanization," 6(1): 69–82 (1976), p. 74, ©1976, Baywood Publishing Co., Inc.

Table 4.2. Comparison of housing unit data for the Northeast Branch Basin 1940 and 1970.

Election District No.	Election District	Housing Units	
		1940	1970
1	Vansville	1,250	6,100
2	Bladensburg	2,663	13,037
10	Laurel	1,684	10,756
14	Bowie	683	7,751
16	Hyattsville	3,804	5,306
17	Chillum	5,128	28,253
19	Riverdale	2,336	7,110
20	Lanham	474	10,413
21	Berwyn	3,005	17,329
5	Colesville	944	14,586
Totals		21,971	120,641

Source: Lazaro, J. Env. Systems, "Planning Aids to Measure the Physical Impact of Urbanization," 6(1): 69–82 (1976), p. 71, ©1976, Baywood Publishing Co., Inc.

Table 4.3. Indicators of imperviousness.

Land-Use Intensity (LIR)	Floor Area Ratio (FAR)	Floor Area (sq ft) per Gross Acre	Density in L.U. per Gross Acre	
			1089 (s.f./L.U.)	871.2 (s.f./L.U.)
0.0	0.0125	544.5	0.5	0.625
1.0	0.025	1089	1.0	1.25
2.0	0.05	2178	2.0	2.5
3.0	0.1	4356	4.0	5.0
4.0	0.2	8712	8.0	10.0
5.0	0.4	17424	16.0	20.0
6.0	0.8	34848	32.0	40.0
7.0	1.6	69696	64.0	80.0
8.0	3.2	139392	128.0	160.0

Source: U.S. Department of Housing and Urban Development (1966), p. 4.

4.5 Estimation of Percent of Imperviousness

The U.S. Department of Housing and Urban Development (1966) has devised (1) the land use intensity ratio (LIR), which refers to the number of living units per unit of land area (acre), (2) the floor area ratio (FAR), which is the amount of floor area (sq. ft.) per unit of area (acre), (3) gross land area, which refers to all of the land which benefits or is used by the housing development (i.e., on-site streets, half-bordering streets), and (4) the gross density, which is the number of housing units divided by the gross land area. Using the various ratios (Table 4.3), a researcher may estimate the percent of imperviousness for a residential region. In order to determine the percent of imperviousness for the remaining portions of the city, one would have to employ aerial photographic methods, as outlined below.

The major shortcoming of the land use intensity method is that it requires an accurate description of the location and size of the areas. Obtaining such data for urbanizing areas which contain a variety of lot sizes may prove to be a very difficult and cumbersome task, at best.

Examination of aerial photographs is perhaps the most accurate method one can employ in deriving estimates of land use (see Section 4.4). Aerial photographs may be carefully planimetered to compute the areas devoted to various land uses. This method, although very accurate, is obviously time-consuming.

Stafford et al. (1974) suggested that the above approach might produce area measurements with a higher degree of accuracy than that of other data used in the particular study, and that a procedure based on sampling would be less laborious and yet adequately meet the accuracy requirements. Sampling methods have been satisfactorily employed by researchers; Wallace (1971) used a sampling density of one point per 20 acres, and Hardy (1970) used a sampling density of one point per 2.5 acres. Gluck and McCuen (1975) utilized a procedure consisting of placing a square grid over aerial photographs at a scale of 1 inch = 400 feet. A hundred point sample was then taken, each point representing approximately 0.9 ac. The predominant land use for the area at the point and immediately around it was listed as the land use for the entire sample square.

Stankowski (1972) proposed the use of population density as an indicator of urban and suburban land use in New Jersey. His formulation was based on correlations between population density and the proportions of land area in each of six urban and suburban land use categories: (1) single-family residential, (2) multiple-family residential, (3) commercial, (4) industrial, (5) public and quasi-public, and (6) conservational, recreational and open. Each category was weighted by its average percentage of impervious area, i.e., low, intermediate or high, and the following equations were derived:

$$I_{low} = 0.170 \, D^{1.165 - 0.094 \log D}$$

$$I_{intermediate} = 0.0218 \, D^{1.206 - 0.100 \log D}$$

$$I_{high} = 0.0263 \, D^{1.247 - 0.108 \log D}$$

where

I_{low}, $I_{intermediate}$ and I_{high} = percentages of impervious land area based on the low, intermediate and high impervious area weighting factors, respectively

D = population density (persons/mi²)

This procedure has one advantage over the use of aerial photographic methods, in that estimates of impervious cover may be obtained for time periods in which aerial photographs are not available. However, the degree of accuracy of estimation of the impervious area directly depends on the careful development of the weighting average.

Gluck and McCuen (1975) suggest that by considering more variables in their prediction, they had improved upon Stankowski's equations (Table 4.4). They supported this statement by carefully planimetering aerial photography of their study area, and comparing results obtained from the equations with the data gathered from the photographs.

Table 4.4. Criteria for determining imperviousness.

Criterion Variable	R	S_e (%)	Model
Impervious Area, %	0.907	6.27	$I = 10.06 + 58.28 \dfrac{0.000128P}{1 + 0.000128P} - 1.258(D - 10.06)$
Residential Housing, %	0.649	5.44	$R = 1.445 P^{0.3150} D^{-0.2688}$
Commercial, %	0.552	1.72	$C = 2.09 P^{0.1045} E^{0.1735} H^{-0.07406} D^{-0.6419}$
Parking, %	0.587	3.59	$PL = 2.725 + 32.70 H^{0.04150} PE^{0.1468} D^{-2.293} + \dfrac{H - 10.0}{11000 - 1.025H}$
Streets, %	0.794	3.23	$S = 15.05 + 19.40 D^{-0.5125} - 115.0 PE^{-0.3147}$
Agricultural, %	0.781	4.05	$AG = -9.4 + 9.2 P^{-0.43} D^{1.21} + \dfrac{E - 25.2}{0.14E - 4}$
Grassland, %	0.460	8.37	$G = 44.09 - 0.00008P + 1.155 \log_e E - 0.00042 \log_e H - 1.188D$
Forest, %	0.788	10.20	$F = 11.21 - 0.00188P - 0.00050E + 0.00080H + 2.297D$

P = population density (persons per square mile)
D = distance (miles)
E = employment density (persons per square mile)
H = housing density (dwelling units per square mile)
A = area (square miles)
PE = population density plus employment density

Source: Gluck and McCuen, *Water Resources Res.*, 11(1):117–178 (1975), p. 179, copyrighted by American Geophysical Union.

Table 4.5. Methods for computing recurrence interval.

California Method:	$T_r = \dfrac{n}{m}$
Hazen Method:	$T_p = \dfrac{2n}{(2m - 1)}$
U.S. Geological Survey Method:	$T = \dfrac{n + 1}{m}$

n = number of years of record
m = the rank starting with the highest as 1
T_p = return period

Source: Dalrymple (1960), pp. 15–16.

4.6 Probabilistic Approaches

Chow (1964) pointed out that:

> . . . one of the most important problems in hydrology deals with interpreting a past record of hydrologic events in terms of future probabilities of occurrence (p. 8-2).

Much of the science of hydrology is based on historical data and, accordingly, probabilistic methods have found widescale application. This is especially true in the study of flooding.

There are two kinds of flood series: the annual flood array, and the partial-duration series (Dalrymple, 1960). The annual flood array is a listing of the largest momentary peak discharge for each year of record. An objection to this method is that in any given year there may be several discharges which exceed annual peaks in other years.

The partial duration series overcomes this discrepancy by establishing a base and considering all floods above this value. The smallest annual flood is generally selected as the base value. Although another criterion for selection is that each event be independent, i.e, be separated by substantial recession of the flood hydrograph, it seems that this may be difficult in practice to obtain, since consecutive flood peaks are often closely spaced in time. In other words, how close is close—where does one draw the line?

Once the type of flood series is selected, there are several methods which can be used to compute the recurrence interval in years (T_r). Some of these are listed in Table 4.5.

It can be argued that each method has its theoretical discrepancies (Dalrymple, 1960). For example, let us take a twenty-year flood series and apply it to all three formulae. By the California method, the largest event is ranked one, and the recurrence interval is twenty years, which is logical. The Hazen method yields a return period of forty years, which represents an artificial extension of record. The U.S. Geological Survey formula gives a recurrence interval of 21 years, which is also logical.

The California method and the U.S.G.S. method are perhaps the most probabilistically sound. In different terminology, the methods state that the highest flood will be equalled 5%, or exceeded 4.76%, of the time. However, the theoretical difference between the two methods becomes obvious when we reverse the ranking order, i.e., list the highest ranked flood event as 20. After computation using the California method, it is found that all floods will be equal to or less than the highest event (but none can exceed it) 100% of the time. The Survey method, however, shows that the highest flood can occur, *be exceeded* or fall short 100% of the time.

URBAN FLOOD FREQUENCY

In general, flood peaks of the shorter recurrence intervals are increased by urbanization, return periods are reduced and a new frequency distribution is established. Researchers have found this phenomenon to be true even for the floods of the longer recurrence intervals. Martens (1968) in his study of metropolitan Charlotte, North Carolina, reported:

> ... the magnitude of the mean annual flood increases with an increase in the degree of imperviousness. The effect of impervious area diminishes with the increased flood recurrence intervals, becoming negligible for floods exceeding 50 years (p. C. 1).

Wilson (1967), after studying flooding in Jackson, Mississippi, drew slightly different conclusions when he found that the mean annual flood for a totally urbanized basin increased by 4.5 times that of a similar rural basin and in addition, the fifty-year flood also increased threefold. Other researchers have derived comparable results to those of Martens and Wilson (see Espey and Winslow, 1974).

Recognizing the shorter lagtimes and higher flood peaks, Carter (1961) developed some regression equations based on an equation presented earlier by Snyder (1958). Carter studied data from watersheds in the Washington, DC area, and considering four assumptions, derived the equations listed below. The assumptions were: (1) the average rainfall/runoff coefficient is 0.3, (2) the effect of the changes in impervious area is independent of the size of the flood, (3) 75% of the rainfall volume on the impervious surface reaches the stream channel, and (4) the impervious area consists of many fairly small areas randomly distributed throughout the basin.

$$K = \frac{0.30 + 0.0045I}{0.30}$$

where K = the factor by which all flood peaks are increased by the percent of impervious area (I)

$$T = 1.20 \left(\frac{L}{\sqrt{S}}\right)^{0.6}$$

where

T = lagtime
L = the total length of the main channel from the gaging point to the rim of the basin
S = the weighted slope in feet per mile of an order of 3 or greater, of all stream channels in the basin

$$\frac{\overline{Q}}{K} 223 A^{0.85} T^{-0.45}$$

where

$\overline{Q}$ = the mean annual flood[34]
A = drainage area in square miles

The latter formula has been used as a model by other researchers (Martens, 1968; Anderson, 1970) in their studies.

CASE STUDY

The author (Lazaro, 1976a) has employed the California method in the presentation of the annual flood series for the Northeast Branch of the Anacostia River, a medium-sized stream (72.8-mi² drainage area) in the Maryland suburbs of Washington, DC. Figure 4.4 shows two curves: curve A is a plot of the entire thirty-two-year annual flood array, and curve B is a plot of the first sixteen years. The x-axis is logarithmic, and the y-axis is linear. Curve B is projected with the hypothesis that if urbanization had not occurred in the watershed, the thirty-two-year curve would more closely approximate the slope of this projection (see Section 4.7 for a discussion of the climatic variables). A 3000 cfs peak can be expected to occur once every three years on curve A, whereas on curve B it would be predicted to happen once every five years. Curve A demonstrates the influence of urban land use change on flood recurrence intervals.

---

[34]Mean annual flood is the flood having a recurrence interval of 2.33 years. See Chow (1964) for a statistical explanation.

This observation may be further strengthened by comparing similar data from a nearby watershed (modification of paired watershed method). Figure 4.5 shows a plot of annual peaks from the Patuxent River above Unity, Maryland, a rural watershed. The data come from a shorter but concurrent time period. Curve A represents a projection of the first half of the record, and curve B is a projection of the total annual flood series. Through the comparison of Figures 4.5 and 4.6, it is suggested that over the study period in the natural, undisturbed state, flood peaks for any given year should have decreased. This is evidenced by curve B in Figure 4.5.

4.7 Statistical Techniques

Inferential statistics may be divided into two groups, parametric and nonparametric (Runyon and Haber, 1968). Essentially, the concepts are the same, i.e., that a sample of a population is statistically tested and assumptions are made concerning the entire population. Parametric tests have very stringent conditions, and any violation of these conditions can lead to erroneous conclusions. McCuen and James (1972) observed the latter, and proposed the use of more lenient nonparametric statistical methods in urban hydrological research. They based their argument on the fact that in many cases, limited data are available on urban areas, and the use of parametric tests would, at best, produce questionable results. However, the researcher must constantly be aware when employing statistical methods that they are based on probability. In cases where human safety is involved (as is often the case in flooding), the researcher should choose the safest design condition and use statistics as *approximations*, and not as absolutes.

CASE STUDY

The author used nonparametric statistical techniques along with conventional methods in a study concerning the effect of urbanization of annual peak flows of the Northeast Branch of the Anacostia River (Lazaro, 1976b). The conventional methods were modifications of those discussed in Section 4.1, and were the paired watershed and before treatment/after treatment methods. These methods were applied to the Northeast Branch watershed, and the Patuxent (see Sections 4.1 and 4.6).

The annual flood series from both basins were examined for a change in central tendency by the Wilcoxon Matched-Pairs Signed-Ranks test. The null hypothesis to be tested was:

H_{null}: There is no difference in the proportion of positive to negative rankings in the two halves of the annual peak series.

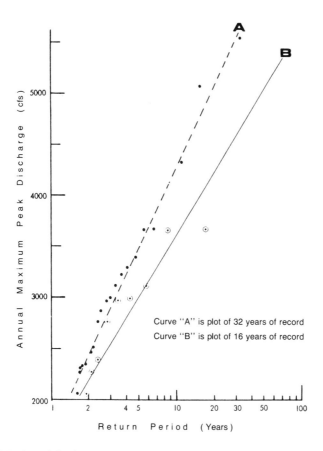

Figure 4.4. Annual flood return period of the Northeast Branch [*Source*: Lazaro, *J. Env. Systems*, "Planning Aids to Measure the Physical Impact of Urbanization," 6(1):69–82 (1976), p. 80, © 1976, Baywood Publishing Co., Inc.].

The period of record for each watershed was divided in half, and each annual peak was listed in descending order in columns 1 and 2 of Tables 4.6 and 4.7. Column 3 lists the differences between these figures. In the last column, the differences are ranked positive if a peak in the second period of record was larger than the first one in the period. The T (statistical) value is the lesser of the sums of the positive and negative signed ranks. A comparison of the T value with a critical value (T_c) determines whether there is a significant difference in central tendency. A T_c value greater than the T value is an indication that most of the larger values correspond to one of the time periods, and thus a difference in central tendency should be expected. With 15 observations, the Northeast Branch T_c is 16; the Patuxent, with 13 observations, has a T_c of 17. Both exceed critical values for significance at 0.01 (Siegel, 1956).

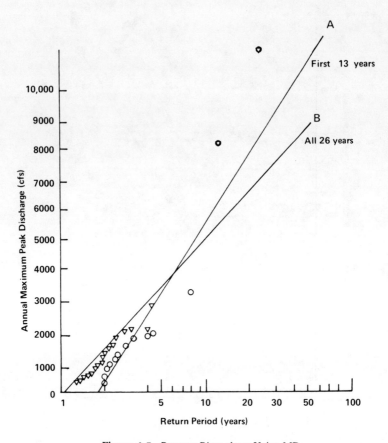

Figure 4.5. Patuxent River above Unity, MD.

An examination of Tables 4.6 and 4.7 reveals $T = 0$ for the Northeast Branch, and $T = 3$ for the Patuxent. Therefore, the null hypothesis can be rejected for both series. There has been a change in central tendency for both annual flood series; however, there was a negative change in the Patuxent data, whereas the change in the Northeast Branch Basin data was positive.

Changing direction of peak central tendency for both watersheds may be explained by a change in (1) the quantity of incoming precipitation, (2) the hydrologic processes within the watersheds, or (3) both. Land use remained unmodified throughout the study period within the Patuxent, and therefore, a change in hydrology is not considered to have occurred. Hence, the decreasing trend in peak central tendency is assumed to be a result of the small rainstorms.

Using the before treatment/after treatment method, rainfall data for the

Table 4.6. Wilcoxon Matched-Pairs Signed-Ranks test applied to the Northeast Branch, annual flood series for water years 1939–1969.

1939–1953 (Qp)[a]	1954–1969 (Qp)[b]	Difference	+ Rank	− Rank
3680	5660	1980	15	
3660	5060	1400	14	
3000	4080	1080	13	
2980	3400	420	1	
2770	3300	530	2	
2350	3240	890	11	
2280	3120	840	9	
2060	2870	810	8	
1980	2830	850	10	
1820	2510	690	5.5	
1780	2470	690	5.5	
1660	2340	680	4	
1350	2310	960	12	
1280	1940	660	3	
889	1670	781	7	
		SUM	=	0
		T	=	0

[a] Qp = maximum annual peak discharge.
[b] There was no recorded peak for the year 1956; therefore, in order to list values for a 15-year period, the peak discharge for the year 1969 was listed.
Modified from Lazaro (1976a), p. 78.

Table 4.7. Wilcoxon Matched-Pairs Signed-Ranks test applied to the Patuxent River above Unity, Maryland, annual flood series for water years 1944–1969.

1944–1956 (Qp)	1957–1969 (Qp)	Difference	+ Rank	− Rank
10700	2920	7780		13
8060	2290	5770		12
3490	1800	1690		11
2220	1590	630		8
2200	1490	710		9
1920	1340	580		7
1830	1090	740		10
1300	934	366		6
1240	920	320		5
1100	828	272		3
1080	788	292		4
494	716	222	2	
240	446	206	1	
		SUM	= 3	
		T	= 3	

Modified from Lazaro (1976a), p. 77.

Northeast Branch were examined. The rainstorms that had produced each of the floods in the annual flood series were listed, and their central tendency tested in a manner similar to Figure 4.2. The results are shown in Table 4.8.

The Wilcoxon Matched-Pairs Signed-Ranks test was applied, and the following null hypothesis was tested:

H_{null}: There is no difference in the proportion of positive to negative rankings in the two halves of the annual rainstorm series.

A negative change in central tendency is observed, significant at the 0.01 level. These data suggest that rainstorms did become smaller. However, peaks were larger. Employment of this nonparametric statistical test suggests that the hydrologic processes (namely, runoff) have been modified within the Northeast Branch Basin. This is demonstrated by the rejection of the null hypothesis at the 0.01 significance levels.

4.8 Urbanized Stream Channels

In his study of a suburbanizing stream in Iowa, Graf (1977) found:

> . . . suburban development adds numerous artificial channels to the previously existing network of natural channels, so that the post development network is partly natural and partly artificial (p. 459).

Figure 4.6 illustrates Graf's observations. Schulz and Lopez (1974) examine the urban surface drainage more closely when they write:

> . . . whereas the country road or urban freeway was built at an elevation above the immediate surroundings, the usual neighborhood street is set at an elevation below the immediate surroundings. This often results in the street functioning as

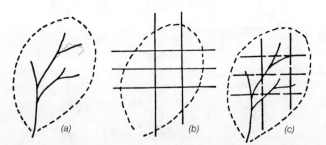

Figure 4.6. Drainage network changes in a hypothetical suburbanizing network. (a) is the original network, (b) is the street, gutter and drain pattern superimposed on the natural pattern by suburbanization, (c) is the total network, partly natural and partly artificial [*Source*: Graf, *Water Resources Res.*, 13(2):459–463 (1977), p. 460, copyrighted by American Geophysical Union].

Table 4.8. *Wilcoxon Matched-Pairs Signed-Ranks test applied to rainstorms producing annual flood series on the Northeast Branch.*

1940–1955	1956–1971	Difference	+ Rank	− Rank
7.7	5.4	−2.3		
7.0	4.6	−2.4		
6.1	4.3	−1.8		
5.2	4.3	−0.9		7.5
5.0	4.1	−0.9		7.5
4.5	3.5	−1.0		9.5
4.3	2.8	−1.5		
3.8	2.6	−1.2		
2.9	2.5	−0.4		5
2.6	2.4	−0.2		2.5
2.3	2.1	−0.2		2.5
2.2	2.1	−0.1		1.5
2.1	2.0	−0.1		1.5
2.0	1.3	−0.7		6
2.0	1.0	−1.0		9.5
		SUM =	0	
		T =	0	

Since the area covered by the basin is comparatively small (72.8 square miles), the arithmetic mean method for determining basin precipitation offered the degree of accuracy needed (Chow, 1964). Values from at least three precipitation stations in and around the watershed were computed and are those listed.
Note: $T = 15$ at a level of 0.01.
Modified from Lazaro (1976a), p. 79.

a drainage way. Usually, there is a crown at the center of the street so that water will drain toward the gutters at either side (p. 11).

Graf summarizes the effects of the altered surface drainage ways when he writes:

> . . . changes in channel networks are such that the network becomes much more efficient in collecting water quickly, so that lag time and kurtosis of storm hydrographs are altered to produce the familiar flash floods of urban areas[35] (p. 459).

Over time, depending upon the climate of a region and the underlying geology, stream channels adjust in volume to accommodate a bankfull discharge, equivalent to a flood with an average recurrence interval of approximately once every 1.5 years (Leopold et al., 1964). Urban land surface changes increase the magnitude and frequency of runoff to a stream, accelerating channel scour and increasing volume. Hammer (1972) investigated these phenom-

---

[35] Graf (1977) in his conclusions warns hydrologists . . . who consider only changes in impervious surfaces risk ignoring an equally serious problem associated with drainage network changes.

ena within seventy-eight small watersheds near Philadelphia. Twenty-eight of the basins contained only rural land uses. Hammer empirically derived the relationship between channel cross-sectional area and watershed area for the rural basins (with a correlation coefficient of 0.87):

$$C = 24.8A^{0.657}$$

where

C = channel cross-sectional area in square feet
A = area of watershed in square miles

The enlargement ratio (R) of an urban stream channel is the proportion of actual channel area to the expected channel area in the absence of urbanization, or:

$$R = C/24.8A^{0.657}$$

Hammer found that the majority of a sample of data from streams in his study area had an enlargement ratio between 1.0 and 2.0.

Leopold (1973) reported the results of a twenty-year study (1953–1972) of river channel changes which were assumed to have occurred as a result of urbanization of Watts Branch, a 3.7-mi^2 watershed in Rockville, Maryland. Leopold noted that although overbank flows increased dramatically (from 2 to over 10 per year), the average cross-sectional area diminished by 20% over the study period.

Hollis and Luckett (1976) reported the results of a stream channel's morphological changes between 1956 and 1970. During the study period, 18% of the Canon Brook watershed in Harlow, Essex, England, was paved. After using a version of the Student's t test (a parametric test), it was concluded that there was no statistically significant change in the morphology of the channel between 1956 and 1970, even though mean monthly floods, summer floods and other floods had increased in magnitude and frequency.

The author has studied cross sections along the tributaries of the Northeast Branch (Lazaro, 1977) and has noted similar irregularities to those observed by the researchers referenced above. It would seem that all of these observations could be summarized by one of Hammer's (1971) statements:

> . . . the effect of urbanization, particularly in the short run, is to increase this variability, (i.e., variability in width, depth, cross-sectional area, etc. over a given channel reach). The scour, deposition, and overall channel enlargement which results from increased peak flows seem to proceed at different rates in different locations (p. 23).

The author completed a preliminary study (Lazaro, 1977) testing the hypothesis that:

> ... a close relationship exists between urban land use change and changes in stream structural parameters, and that changes in stream structural parameters directly reflect the much larger surface structural changes occurring on the remaining watershed. Hence, stream structural parameters may be used as gauges to define what is occurring, on a much larger scale, in the watershed itself (p. 322).

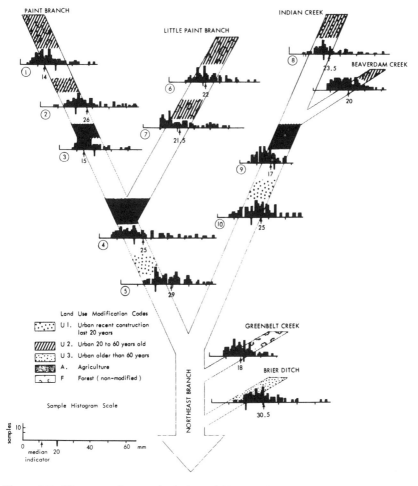

Figure 4.7. Histograms of coarse riverbed material intermediate axes [*Source*: Lazaro, *J. Env. Systems*, "Adapted-Techniques for Urban Stream Structure Analysis," 6(4):321–328, p. 325, © 1977, Baywood Publishing Co., Inc.].

Table 4.9. Northeast Branch tributary cross section areas.

Stream Cross Section	Drainage Area (Above Cross Section) (sq mi)[a]	Cross-Sectional Area (sq. t)[b]
Paint Branch (1)	6.7	431
Paint Branch (2)	10.5	429
Paint Branch (3)	13.3	142
Paint Branch (4)	23.3	356
Paint Branch (5)	26.2	330
Little Paint Branch (6)	7.0	80
Little Paint Branch (7)	8.3	130
Beaverdam Creek	6.5	60
Indian Creek (8)	6.8	76
Indian Creek (9)	23.0	141
Indian Creek (10)	26.7	288
Brier Ditch	3.9	312
Greenbelt Creek	2.9	77

[a] J. O. Duru, Storm Water Management Study: Northeast-Northwest Branch, Anacostia River. Prince Georges County, Department of Planning, September, 1973. Figure 5.
[b] Computed by author by forming a trapezoid.
Source: Lazaro, J. Env. Systems, "Adapted Techniques for Urban Stream Structure Analysis," 6(4):321–328 (1977), p. 327, ©1977, Baywood Publishing Co., Inc.

The Northeast Branch Basin (described in Sections 4.4 and 4.6) was divided into six sub-watershed areas, which are schematically shown in Figure 4.7. Some of these areas were used as controls. Not only were cross-sectional area data collected, but also the median and distribution of coarse riverbed material,[36] using methods modified from Wolman (1954) (see also Lazaro, 1978).[37] Thirteen sites were selected, following certain criteria. Cross sections were computed, the intermediate axes of 100 particles were measured, and a histogram plotted for each site. Land use upstream from each cross section was noted. Figure 4.7 and Table 4.9 display all these data.

The author arrived at the following conclusions:

(1) Data collected from drainage areas with rural and a mixture of rural-urban land use modifications displayed smaller histogram dispersions, smaller medians and increased cross-sectional areas as drainage area increased.
(2) Data from Brier Ditch, the oldest urbanized area studied, displayed the largest dispersion, the largest median and a cross-sectional area that was very large for the area drained, when compared to Greenbelt Creek.

These indicators suggest that the degree and intensity of urban land use modifications are directly reflected in the variability of the channel data. The

---
[36] Coarse riverbed particles are the gravel-to-cobble sized material found on the streambed.
[37] In this paper, the author describes in detail the methods of data collection and analysis.

more recent and intense the land use modification, the wider the histogram dispersion, the larger the median of the coarse riverbed intermediate axis, and the more irregularly defined the tributary cross-sectional areas.

4.9 Summary

This book was written with the intention of providing a sound basis for further urban hydrologic studies. The referenced sources are considered to be the major sources and are recommended for use by the reader.

An urban hydrologic study in scope and purpose is similar to other studies. One must categorize the study (quantity or quality), scale it by considering data available, time and cost limitations, and justify expenses for product derived. Census data collected and enumerated by the U.S. Department of Commerce, Bureau of the Census, streamflow and water quality data collected and compiled by the U.S. Geological Survey, Water Resources Division, or other monitoring agencies, and climatic data collected and compiled by the U.S. Weather Bureau are usually the data required for an urban hydrologic study. These data may be found in large college libraries, or many offices of the U.S. Geological Survey, Water Resources Division located all over the United States. Land use data may be found at many planning commissions or may be obtained by the researcher using aerial photography of formulae presented in Section 4.5.

Primary data collection refers to data collected by the researcher. However, it must be cautioned that data collection can be hazardous and that the researcher should take every precaution to ensure personal safety. Methods of collecting streamflow data and water quality data are outlined.

Graphic means of displaying population density changes and shifts are presented. The problem here is that of obtaining an accurate estimate of the watershed population. Aerial photography may provide a visual means of displaying population density and land use shapes accompanying urbanization on smaller watersheds.

Several methods of estimating the percent of imperviousness are presented. Examination of aerial photography provides the most accurate method; however, it may also be very time-consuming. Stankowski (1972) developed equations relating imperviousness to population density. Gluck and McCuen (1975) improved upon the work by Stankowski, and formulated several equations for estimating imperviousness due to different land uses.

Much of the science of hydrology is based on historical data and, accordingly, probabilistic methods have found wide-scale application. These methods supply the fundamentals for the study of flooding. Flood series, as outlined by Dalrymple (1960), are presented. Urbanization tends to increase flood peaks of shorter recurrence intervals (Martens, 1968). Wilson (1967) not only showed this, but additionally suggested that fifty-year floods were also increased. Several regression equations are presented which attempt to describe the expected change due to urban land surface alterations. The writer demonstrates methods he employed in graphically displaying the changes in recurrence intervals following urbanization (Lazaro, 1976a).

Nonparametric statistical techniques have been suggested for use in urban hydrology by McCuen and James (1972). The author (Lazaro, 1976b) used conventional methods (paired watershed and before/after) along with a nonparametric statistical test to demonstrate the change in flood peaks accompanying urbanization. These methods are easy to employ, and may yield clear indications of the changes in runoff processes.

When the land surface structure is modified by urbanization, the natural drainage patterns are also altered creating a new drainage density. The effect of these modifications is to cause the urban stream channel to enlarge (Hammer, 1972) in an irregular fashion. Several researchers have studied this phenomenon (Leopold, 1973; Hollis and Luckett, 1976; Hammer, 1972; Lazaro, 1977). The writer presents results of a preliminary study he completed on an urbanizing watershed. In general, the results obtained displayed great variability.

4.10 Exercises

1 Write up an outline for an urban hydrologic study in which you will examine over a ten-year period the change in imperviousness as it relates to the increase in amplitudes of peak flows at a selected point on an urban stream.
   a) What data will you need? Will you collect them yourself?
   b) How will you determine the change in the impervious area?
   c) What method will you use to determine changes in peak flows? Why?

2 Write up the outline for an urban hydrologic study in which you will examine, over a ten-year period, the water quality parameter which you believe to have experienced the greatest change in concentration at a selected point on an urban stream.
   a) What evidence do you have to support your choice of this parameter?
   b) What data will you need? Will you collect them yourself? What method will you use to preserve the quality of the samples?

References

ALLEY, W. M. "Guide for Collection, Analysis and Use of Urban Stormwater Data," ASCE Conference Report, November 28–December 3, 1976, 115 pp. (1976).

References

ANDERSON, D. G. "Effects of Urban Development on Floods in Northern Virginia," U.S. Geological Survey Water Supply Paper No. 2001-C, 22 pp. (1970).

BEST, G. S. and S. L. Ross. *River Pollution Studies*. Liverpool, England:Liverpool University Press, 92 pp. (1977).

BROWN, E. et al. "Methods for Collection and Analysis of Water Samples for Dissolved Minerals and Gases," U.S. Geological Survey, Techniques for Water Resources Investigations, Chapter A1, Book 5, 160 pp. (1970).

BUCHANAN, T. J. and W. P. Somers. "Stage Measurement at Gaging Stations," U.S. Geological Survey Techniques for Water Resources Investigations, Book 3, Chapter A7, 28 pp. (1968).

BUCHANAN, T. J. and W. P. Somers. "Discharge Measurements at Gaging Stations," U.S. Geological Survey Techniques for Water Resources Investigations, Chapter A8, Book 3, 65 pp. (1969).

CARTER, R. W. "Magnitude and Frequency of Floods in Suburban Areas," U.S. Geological Survey Professional Paper No. 424-B, 11 pp. (1961).

CHAPMAN, T. G. and F. X. Dunin. *Prediction in Catchment Hydrology*. Australian Academy of Science, Netley, Australia:The Griffin Press, 498 pp. (1975).

CHORLEY, R. J., ed. *Water, Earth and Man*. London:Metheun and Co., Ltd., 588 pp. (1969).

CHOW, V. T., ed. *Handbook of Applied Hydrology*. New York:McGraw-Hill Book Company (1964).

CIACCIO, L. L., ed. *Water and Water Pollution Handbook, Vol. 1*. New York:Marcel Dekker, Inc., 449 pp. (1971).

DALRYMPLE, T. "Flood Frequency Analyses: Manual of Hydrology, Part 3, Flood Flow Techniques," U.S. Geological Survey Water Supply Paper No. 1543-A (1960).

DETWYLER, T. R. and M. Marcus. *Urbanization and the Environment*. Belmont, CA:Duxbury Press, 287 pp. (1972).

ESPEY, W. H., Jr. and D. E. Winslow. "Urban Flood Frequency Characteristics," *J. Hyd. Div.*, ASCE, 100(HY2):279–293 (1974).

GREESON, P. E. et al. "Methods for Collection and Analysis of Aquatic Biological and Microbiological Samples," U.S. Geological Survey, Techniques of Water-Resources Investigations, Chapter A4, Book 5, 332 pp. (1977).

GLUCK, W. R. and R. H. McCuen. "Estimating Land Use Characteristics for Hydrologic Models," *Water Resources Res.*, 11(1):177–179 (1975).

GRAF, W. L. "Network Characteristics in Suburbanizing Streams," *Water Resources Res.*, 13(2):459–463 (1977).

GREGORY, K. J. and D. E. Walling. *Drainage Basin Form and Process: A Geomorphological Approach*. London:Edward Arnold Publishers, 456 pp. (1973).

GUY, H. P. and V. W. Norman. "Field Methods for Measurement of Fluvial Sediment," U.S. Geological Survey Techniques of Water-Resource Investigations, Chapter C2, Book 3, 59 pp. (1970).

HAMMER, T. R. "The Effect of Urbanization on Stream Channel Enlargement," Ph.D. Dissertation, The University of Pennsylvania, Philadelphia, PA (1971).

HAMMER, T. R. "Stream Channel Enlargement Due to Urbanization," *Water Resources Res.*, 8(6):1530–1540 (1972).

HARDY, E. E. "Inventorying New York's Land Use and Natural Resources," *N. Y. Food Life Sci.*, 3(4) (1970).

HARRIS, E. E. and S. E. Rantz. "Effect of Urban Growth on Streamflow Regimen of Parmanente

Creek, Santa Clara County, California," U.S. Geological Survey, Water Supply Paper No. 1591-B, 18 pp. (1964).

HOLLIS, G. E. and J. R. Luckett. "The Response of Natural River Channels to Urbanization: Two Case Studies from Southeast England," *J. Hydrology*, 30(4):351-363 (1976).

KITTRELL, F. W. "A Practical Guide to Water Quality Studies of Streams," U.S. Dept. of the Interior, Federal Water Pollution Control Administration, CWR-5, 135 pp. (1969).

LAZARO, T. R. "Planning Aids to Measure the Physical Impact of Urbanization," *J. Env. Systems*, 6(1):69-82 (1976a).

LAZARO, T. R. "Nonparametric Statistical Analysis of Annual Peak Flow Data from a Recently Urbanized Watershed," *Water Resources Bull.*, 12(1):101-107 (1976b).

LAZARO, T. R. "Adapted Techniques for Urban Stream Structure Analysis," *J. Env. Systems*, 6(4): 321-328 (1977).

LAZARO, T. R. "Urban Land Use and Its Impact on Streams: A Field Exercise," *Sci. Education*, 62(2):165-171 (1978).

LEOPOLD, L. B. "Hydrology for Urban Land Planning—A Guidebook on Hydrologic Effects of Urban Land Use," U.S. Geological Survey Circular No. 554, 18 pp. (1968).

LEOPOLD, L. B. "River Channel Change with Time: An Example," *Geol. Soc. Am. Bull.*, 84: 1845-1860 (1973).

LEOPOLD, L. B. et al. *Fluvial Processes in Geomorphology*. San Francisco:W. H. Freeman & Company, Publishers, 522 pp. (1964).

LINSLEY, R. K., Jr. et al. *Applied Hydrology*. New York:McGraw-Hill Book Co., 689 pp. (1949).

MCCUEN, R. H. and L. D. James. "Nonparametric Statistical Methods in Urban Hydrologic Research," *Water Resources Bull.*, 8(5):965-975 (1972).

MARTENS, L. A. "Flood Inundation and Effects of Urbanization in Metropolitan Charlotte, N.C.," U.S. Geological Survey Water Supply Paper No. 1591-C, 60 pp. (1968).

RUNYON, R. P. and A. Haber. *Fundamentals of Behavioral Statistics*. Reading, MA:Addison-Wesley Publishing Co., Inc., 304 pp. (1968).

SCHULZ, E. F. and O. G. Lopez. "Determination of Urban Watershed Response Time," Hydrology Paper No. 71, Colorado State University, Fort Collins, CO, 41 pp. (1974).

SIEGEL, S. *Nonparametric Statistics for the Behavioral Sciences*. New York:McGraw-Hill Book Co., 312 pp. (1956).

SLACK, K. V. et al. "Methods for Collection and Analysis of Aquatic Biological and Microbiological Samples," U.S. Geological Survey Techniques in Water-Resources Investigations, Chapter A4, Book 5, 165 pp. (1973).

SNYDER, F. F. "Synthetic Flood Frequency," *J. Hyd. Div.*, ASCE Paper 1808 (HY5):1-22 (1958).

SNYDER, W. M. "Summary and Evaluation of Methods for Detecting Hydrologic Changes," *J. Hydrology*, 12:311-338 (1971).

STAFFORD, D. B. et al. "Use of Aerial Photographs to Measure Land Use Changes," paper presented at the National Meeting on Water Resources Engineering, American Society of Civil Engineers, Los Angeles, California (1974).

STANKOWSKI, S. J. "Population Density as an Indirect Indicator of Urban and Suburban Land-Surface Modifications," U.S. Geological Survey Professional paper No. 800-B, 25 pp. (1972).

TIMBERMAN, C. W. "Discussion of Some Aspects of Urban Hydrology Methodology," Paper No. 7 in *U.S. Army Corps of Engineers, Proceedings of a Seminar on Urban Hydrology*, September 1-3, 1970, Davis, California, 73 pp. (1970).

U.S. BUREAU OF CENSUS, CENSUS OF POPULATION. "General Population Characteristics," Final Report PC(1)-B1 U.S. Summary (1970).

U.S. DEPARTMENT OF HOUSING AND URBAN DEVELOPMENT. "Land Use Intensity," Land Planning Bulletin No. 7, 10 pp. (1966).

U.S. GEOLOGICAL SURVEY, OFFICE OF WATER DATA COORDINATION. *National Handbook of Recommended Methods for Water-Data Acquisition.* Chapter 5 (1977).

VELZ, C. J. *Applied Stream Sanitation.* New York:John Wiley and Sons, Inc., 619 pp. (1970).

VIESSMAN, W., Jr. et al. *Introduction to Hydrology.* New York:Harper and Row, Publishers, Inc., 704 pp. (1977).

WALLACE, J. R. "The Effects of Land Use Change on the Hydrology of an Urban Watershed," Report No. ERC-0871, School of Civil Engineering, Georgia Institute of Technology, Atlanta, GA, 66 pp. (1971).

WILSON, K. V. "A Preliminary Study of the Effect of Urbanization on Floods in Jackson, Mississippi," U.S. Geological Survey Professional Paper No. 575-D, pp. 259–261 (1967).

WOLMAN, M. G. "A Method of Sampling Coarse Riverbed Material," *Trans. Am. Geophysical Union*, 35(6):951–956 (1954).

Selected Readings

Barnes, I. "Field Measurement of Alkalinity and pH," U.S. Geological Survey, Water Supply Paper 1535-H, 17 pp. (1964).

Beard, L. R. "Probability Estimates Based on Small Normal Distribution Samples," *J. Geophys. Res.*, 65(7):2143–2148 (1960).

Benson, M. A. "Evaluation of Methods for Evaluating the Occurrence of Floods," U.S. Geological Survey Water Supply Paper 1580-A, 30 pp. (1962).

Clarke, R. T. "Statistical Methods for the Study of Spatial Variations in Hydrological Variables," in *Facets of Hydrology.* J. C. Rodda, ed. London:John Wiley and Sons, pp. 299–314 (1976).

Colby, B. R. "Relationship of Sediment Discharge to Streamflow," U.S. Geological Survey Open File Report, 170 pp. (1956).

Cyrus William Rice Division, NUS Corporation. "Design of Water Quality Surveillance Systems: Phase I—Systems Analysis Framework," prepared for the Federal Water Quality Administration, Department of the Interior, Program No. 16090 DBJ, Contract No. 14-12-476, 303 pp. (1970).

Emerson, J. W. "Channelization: A Case Study," *Science*, 173:325–326 (1971).

Fleming, G. and D. D. Franz. "Flood Frequency Estimating Techniques for Small Watersheds," *J. Hyd. Div.*, ASCE, 97(HY9):1441–1460 (1971).

Gann, E. E. "Generalized Flood-Frequency Estimate for Urban Areas in Missouri," U.S. Geological Survey Open File Report, 18 pp. (1971).

Goerlitz, D. F. and E. Brown. "Methods for Analysis of Organic Substances in Water," in *Techniques for Water-Resources Investigations*, U.S. Geological Survey, Chapter A3, Book 5, 40 pp. (1972).

Hardison, C. "Accuracy of Streamflow Characteristics," U.S. Geological Survey Professional Paper 650-D, pp. D210–D214 (1969).

Hardison, C. and M. Jennings. "Bias in Computed Flood Risk," *J. Hydr. Div.*, ASCE 98(HY3):415–427 (1972).

Harrold, L. L. "Estimating Flood Volumes and Hydrographs Corresponding to Peak Flows of Given Frequencies for Small Agricultural Watersheds," *J. Geophys. Res.*, 67(11):4341–4346 (1962).

Hordon, R. M. "Changing Watersheds in a Metropolitan Environment: A Statistical Analysis of Selected Basins in Northeastern New Jersey," in *Watersheds in Transition.* S. C. Csallany et al., eds. Urbana, IL:American Water Resources Association, pp. 394–399 (1972).

Hultquist, N. B. "Water Quality and Quantity as an Aspect of Dynamic Urbanism," Institute of Urban and Regional Research, University of Iowa, Technical Report No. 4, 30 pp. (1972).

Hydrologic Engineering Center, U.S. Army Corps of Engineers. *Proceedings of a Seminar on Urban Hydrology,* September 1–3, 1970, Davis, California, 73 pp. (1970).

Ingram, M. I. et al. "Biological Field Investigative Data for Water Pollution Surveys," U.S. Department of the Interior, Federal Water Pollution Control Administration, No. WP-13, 139 pp. (1966).

Interagency Committee on Water Resources. "A Study of Methods Used in Measurement and Analysis of Sediment Loads in Streams," *Determination of Fluvial Sediment Discharge,* Report No. 14, 151 pp. (1963).

Jennings, M. E. and M. A. Benson. "Frequency Curves for Annual Flood Series with Some Zero Events of Incomplete Data," *Water Resources Res.,* 5(1):276–280 (1969).

Johnson, F. A. "A Note on River Sampling and Testing," *Water & Water Eng.,* 75:59–61 (1971).

Kisiel, C. C. "Time Series Analysis of Hydrologic Data," in *Advances in Hydroscience, Vol. 5.* V. T. Chow, ed. New York:Academic Press, Inc., pp. 1–119 (1969).

Krizek, R. J. and E. F. Mosonyi, eds. *Water Resources Instrumentation, Vol. 1: Measuring and Sensing Methods.* Ann Arbor, MI:Ann Arbor Science Publishers, Inc., 596 pp. (1974).

Langbein, W. B. "Annual Floods and the Partial Duration Series," *Trans. Am. Geophysical Union,* 30(6):879–881 (1949).

Lara, O. G. "Effects of Urban Development on the Flood-Flow Characteristics of the Walnut Creek Basin, Des Moines Metropolitan Area, Iowa," U.S. Geological Survey, Water Resources Investigations, 78-11, 31 pp. (1978).

Lehmann, E. L. and H. J. M. D'Abrera. *Nonparametric Statistical Methods Based on Ranks.* San Francisco:Holden-Day Inc. (1975).

Leopold, L. B. and H. E. Skibitzke. "Observations on Unmeasured Rivers," *Geog. Ann.,* 49A:247–255 (1967).

McCuen, R. "A Comparison of Methods of Estimating Impervious Areas," paper presented at the Spring Annual meeting, American Geophysical Union, Washington, DC, June, 1975. Abstract in EOS, *Trans. Am. Geophysical Union,* 56:361 (1975).

McCuen, R. and H. W. Piper. "Hydrologic Impact of Planned Unit Developments," *J. Urban Planning and Dev. Div.,* ASCE, 101(UPI):93–102 (1975).

Maddock, T., Jr. "Indeterminate Hydraulics of Alluvial Channels," *J. Hyd. Div.,* ASCE, (HY11):2309–2323 (1970).

Matalas, N. C. "Time Series Analysis," *Water Resources Res.,* 3(3):817–829 (1967).

Melton, M. A. "Methods for Measuring the Effect of Environmental Factors on Channel Properties," *J. Geophys. Res.,* 67(4):1485–1490 (1962).

Morel-Seytoux, H. J. and F. Saheli. "A System Approach to Minimal Time Detection of Changes in Watersheds," in *Watersheds in Transition.* S. C. Csallany, et al., eds. Urban, IL:American Water Resources Association, pp. 124–131 (1972).

Needham, J. G. and P. R. Needham. *A Guide to the Study of Fresh-Water Biology.* San Francisco:Holden-Day, Inc., 109 pp. (1962).

Rainwater, F. H. and L. L. Thatcher. "Methods for Collection and Analysis of Water Samples," U.S. Geological Survey Water Supply Paper 1454, 301 pp. (1960).

Rantz, S. E. "Urban Sprawl and Flooding in Southern California," U.S. Geological Survey Circular No. 601-B, 11 pp. (1970).

Reimer, P. O. and J. B. Franzini. "Urbanizations' Drainage Consequences," *J. Urban Planning & Dev.*, ASCE Paper 8600, 97(UP2) (1971).

Riggs, H. C. "Frequency of Natural Events," *J. Hyd. Div.*, ASCE (HY1):15–26 (1961).

U.S. Water Resources Council. "Guidelines for Determining Flood Flow Frequency," Bulletin No. 17A, June, 1977, Washington, DC (1977).

Vemuri, V. and N. Vemuri. "On the Systems Approach in Hydrology," *Bull. Int. Assoc. Sci. Hydrol.*, XV(2):17–38 (1970).

Water Resources Council, Hydrology Committee. "A Uniform Technique for Determining Flood Flow Frequencies," Bulletin No. 15, Washington, DC (1967).

Yevjevich, V. *Probability and Statistics in Hydrology*. Fort Collins, CO:Water Resources Publications, 302 pp. (1972).

5 | Modeling Urban Water Quantity and Quality

... the only perfect model of a watershed is, of course, the watershed itself. The objective of any mathematical model is to tranform the natural geometry into a simpler geometry and yet retain a similar hydrological response. — Singh and Woolhiser (1976, p. 222).

5.1 Mathematical Models

SNYDER AND STALL (1965) define a model as:

> ... simply the symbolic form in which a physical principle is expressed. It is an equation or formula, but with the extremely important distinction that it was built by consideration of the pertinent physical principles, operated by logic, and modified by experimental judgement and plain intuition (p. 87).

Andrews and McLone (1976) expanded upon the latter when they stated:

> ... mathematical modeling is the art of applying mathematics to a real-life situation. A good model recognizes the relevant features of a problem by means of a judicious choice of assumptions and has a well-defined mathematical structure from which the quantities of practical interest can be derived (back cover).

Roberts (1976) presented mathematical model building as being a four-stage cycle. The translation from real-world data to the conceptualization of the model proceeds by induction, i.e., a general law is guessed at on the basis of several observations. Predictions from the model are accomplished through deduction; conclusions are derived on the basis of specified assumptions and well-known rules of inference. The third step concerns the decision on how accurate the predictions are in comparison to the real-world data. This step, as Roberts points out, may require a critical decision from the model designer, since it depends on the use to which the model will be put. The last step is testing the model results against the real-world data and consequent refinement of the model predictions.

Figure 5.1 illustrates the modeling process:[38] (1) the real-life situation is ob-

---

[38] For a comprehensive discussion of models and systems, see Chorley and Kennedy (1971).

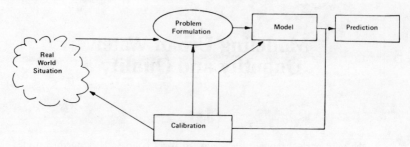

Figure 5.1. The modeling process [modified from: McLone (1976)].

served, (2) the problem is isolated and appropriate concepts are developed, (3) the mathematical model is structured, (4) the model is tested, curve fitted, optimized, etc., and (5) the model is used to make several predictions and further calibrate and validate or "fine tune" the model parameters (sensitivity is improved).

DEVELOPMENT OF A HYPOTHETICAL LINEAR MODEL

Real-World Situation — A metal drum with a capacity of 100 gal is located directly under a spout which drains the roof of a small shed. There is a rain gauge nearby. The drum fills a little after each rainstorm. Construct a mathematical model which will define this system.[39]

First Step — The Problem Formulation — Input is supplied by precipitation, and output consists of the volume of water collected by the roof and deposited in the barrel.

Second Step — Model Structuring — To develop a relationship between the inflow and the increase in water volume observed in the barrel, we will plot a scatter diagram. We observe several rainstorms, and after carefully reading the rain gauge and measuring the increases in volume, we plot these data on Figure 5.2. Spiegel (1975) demonstrates the method of least squares which we can use to fit a curve to these data points. This method is used and best-fit line on Figure 5.2 is drawn.

Since the relationship of rainfall to volume increases in this case and is represented by a straight line and the x and y axes in the figure are in linear increments, then we can assume that a linear relationship exists between rain-

---

[39]System is defined after Dooge (1968) as ". . . any structure, device, scheme or procedure, real or abstract, that interrelates in a given time reference, an input, cause or stimulus, of matter, energy or information, and an output effect or response of information, energy or matter" (p. 58).

fall and volume produced, and our mathematical model must be on the order of:

$$y = a + bx$$

Substituting:

$$R = a + bV$$

where

R = measured rainfall in inches (independent variable)
V = measured volume increase in gallons (dependent variable)
a = a constant
b = slope

Using methods demonstrated by Spiegel (1975), we obtain the coefficients and arrive at:

$$R = a + 10V \; (a = 0 \text{ and } b = 1); \text{ and } R = 10V \quad (5.1)$$

or 1 inch of rainfall produces a 10-gal increase in water volume.

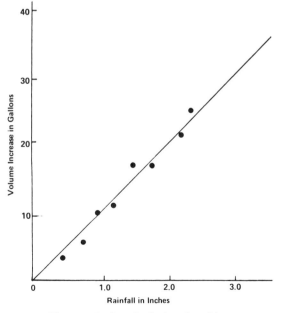

Figure 5.2. Dataplot for barrel model.

Third Step—Testing of Model and Curve Fitting[40]—After a storm, we read the rain gauge and find that 0.5 in. have fallen. By examining Figure 5.2 for this value, we predict that there is a 5-gal increase in the drum. However, after observing the volume in the barrel, we find that there is 3.8-gal increase. If we plotted this point on the figure, we would find that it would be the third point to fall below the regression line or, in other words, the data indicate that the model seems to have nonlinear relationships below 1 inch of rainfall. Here we can raise these points (cause them to track the curve more closely) by changing our coefficient b from 10 to something higher, say 13. If we multiply 3.8 by 13.2, we get 5.02 gal, and this point plots just a little above the curve.

If we observe several rainstorms, we might find other nonlinearities occurring which can be brought closer to the curve by curve fitting (optimization). These latter points may be called outliers, and in these cases are probably caused by the roof of the shed being hot and evaporating a portion of the incoming rainfall before runoff begins. If we further investigate the other outliers, we will most likely find other meteorological explanations for their nonlinear behavior.

In Equation (5.1), the a constant might have to be assigned a value and b might have to have different values for different rainfall magnitudes and intensities. These values for a and b could be placed in tabular form according to the range of rainfall.

DISCUSSION OF MODELING APPROACHES

The model which has just been developed may be called a simple (parametric) rainfall/runoff (volume) regression model (Pilgrim 1975). It is essentially a "black box" where the output is functionally related to the input (Dunin 1975).[41]

Had we tried to portray and predict mathematically the various processes of rainfall and runoff by determining the imperviousness of the roof, the hydraulic efficiency of the gutter network, the intensity and distribution of rainfall, we could call this new model "deterministic." Overton and Meadows (1976) classify modeling in three approaches: (1) deterministic, (2) parametric, and (3) purely random (stochastic) (Figure 5.3). A deterministic system is one which has no element of chance in it; hence, for a given input, the output may be predicted. The difference between deterministic and parametric is a matter of degree, and in the focus, as Overton and Meadows write:

---

[40]It is not possible to calibrate this model given the stated conditions, since no other (more accurate) data are available. Calibration in a gauged watershed would consist of comparing the model predictions against gauged data and modifying the model output to more closely track the real data.

[41]See Clarke (1973) for a thorough discussion of hydrologic models and their classification.

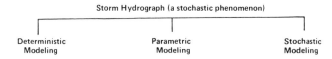

Figure 5.3. Modeling methods [modified from: Overton and Meadows (1976)].

... the parametric approach strives for the definition of the functional relations between hydrologic and geometric and land use characteristics of a catchment (p. 159).

The deterministic approach strives to portray comprehensively (mathematically) the various processes in achieving a very accurate representation of the physical hydrologic system. A stochastic model has probability associated with output.

5.2 Modeling a Natural Watershed

Hydrologists are concerned with modeling the surface and subsurface processes of the hydrologic cycle within a particular watershed (Dooge, 1968). Figure 5.4 illustrates in block-diagram form the various components of a catchment as a closed system. Conceptually and qualitatively, this system and its component processes may be described quite easily. However, the development of accurate mathematical models is complicated by at least one major factor. As is characteristic of many real-world systems, the hydrologic system is nonlinear,[42] as was pointed out by Amorocho (1963, 1967a, 1967b) and Prasad (1965). Amorocho (1967a) classified the non-linear prediction problem as involving:

(1) Time variability of watersheds

(2) Uncertainty with respect to the time and space distribution of the inputs of hydrologic systems

(3) Inherent nonlinearity of the processes of mass and energy transfer that constitute the hydrologic cycle

Time variability may be explained by the fact that a natural watershed is continuously changing over geologic time. Therefore, an input at a particular time is not likely going to produce the same exact output as a similar input later on in time. Our concern is with predicting this input–output relationship over a relatively brief amount of time, but in essence we are plagued by similar problems. These could be classified as seasonal changes such as trees in leaf, a permeable ground in warm weather versus an impermeable ground in

---

[42]For an excellent discussion of nonlinear as well as linear systems, see Cosgriff (1958).

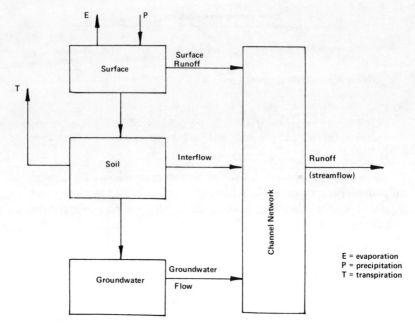

Figure 5.4. Catchment as a closed system [modified from: Dooge (1968)].

winter, precipitation in liquid or solid form, etc. These factors vary the lagtime at any given instant, and render it unpredictable. In addition, a watershed has memory, i.e., the effects of a large flood will be observable in and along the stream channels for a period of time, depending on the magnitude of the flood event and the geology of the region (Wolman and Miller, 1960; Gupta and Fox, 1974). These morphological changes will modify watershed response and consequent outputs (add to nonlinearity) from storms during this time period.

System uncertainties with respect to the space and time distribution of inputs and outputs are largely explained by the inefficiencies of present data collection methods. For example, it is well-known that precipitation at any given instant is not uniformly distributed across a catchment. Rain gauges are the most accurate form of measurement; however, they leave much to be desired in that they measure point rainfall and give accurate estimates for only a small surrounding area, leaving huge, uncharted voids in the storm field. Soil moisture concentrations and depth measurements suffer much the same fate. In short, data from each subsystem of Figure 5.2 add their own nonlinearities to predictive efforts.

Point three, inherent nonlinearity of the processes of mass and energy transfer that constitute the hydrologic cycle, is essentially similar to what has just been presented, with the sole difference being that the scale has been expanded to include the globe.

Thus, the problem of prediction of natural watershed outputs with any given input is a complex one. Hydrologists try to overcome these barriers by making assumptions. As Dooge (1968) reported:

> ... if no assumptions are made about the nature of the system, then the problem of prediction is virtually *insoluble*. The more assumptions that are made about the system, the easier becomes the solution of the problem, but the greater the risk of the failure of our model system to (accurately) reflect the prototype (p. 63).

There are numerous watershed models in use today. Some, if not all, make the following assumptions: (1) time invariance; (2) linearity of the hydrologic subsystems; (3) lumped parameters, i.e., assuming that the parameter for rainfall, infiltration, interflow, etc., is representative of an "average" or net effect of the respective process over the entire catchment (Huggins and Monke, 1968).

Some modeling approaches have attempted different (innovative) means, with varying degrees of success, to overcome prediction problems. Reed, Johnson and Firth (1975) tried to solve the nonlinearities introduced by the variable time lag by inputting their model with a sequence of successive elements of rainfall excess (rainfall occurring after surface saturation has been accomplished) of equal duration. Each such element is distributed with a change in time hydrograph to produce an elemental runoff layer (also see Jacoby, 1966).

Chiu and Bittler (1969) suggested that nonlinearities produced by infiltration could be removed from the rainfall–runoff relationship, and a linear system created (they introduced the concept of a "linear" time-varying model). They suggested that rainfall excess should be the input, instead of observed rainfall data, and that baseflow should be removed from total runoff in development of a unit hydrograph. Chiu and Hwang (1970) further modified these concepts to design a nonlinear time-varying model.

Huggins and Monke (1968) developed a model which avoided the use of lumped parameters by delineating the watershed as a grid of small, independent elements. In this manner, the significant physical hydrologic parameters of slope, infiltration rates, etc., are more precisely accounted for. The major problem with this model is that it requires data inputs which are not as yet readily available. Each approach has its advantages and disadvantages when examined closely.

5.3 Rainfall–Runoff Modeling

The number of working models which may be directly employed (or easily modified) to predict urban runoff probably number well over a hundred. All of these models, no matter how complex they may seem, have one or more of the "basic building blocks," or components (Dawdy, 1969) incorporated within their structure, and usually their differences lie in the comprehensiveness of the portrayal and deployment of one or more of these blocks. This section will briefly discuss these components and present some of the limitations which any runoff modeler may encounter.

Singh and Birsoy (1977) state:

> . . . the relationship between rainfall and runoff is one of the most important problems in hydrology. It is also one of the most difficult problems. The rainfall–runoff relationship quantifies the response function describing the behavior of a watershed. The response function is a result of numerous processes, complex and interdependent, that participate in the transformation of rainfall into runoff. The transformation process encompasses virtually the entire domain of [the] hydrologic cycle. This all-encompassing nature of the transformation process is largely responsible for the complexity underlying the rainfall–runoff relationship. The complexity is enhanced further by spatial and temporal variability of hydrometeorological conditions, and watershed physiography, as well as their interacting influences (p. 251).

In Figure 5.5, the relationship between storm rainfall and flood runoff is divided into three response subsystems (Dooge 1968). Mein and Larson (1973) report that for the continental United States, greater than 70% of annual precipitation infiltrates into the soil. Even though this may be very true, many model designers have chosen to ignore the effects of interflow, soil moisture and groundwater flow, and to model mathematically the surface runoff response. The former is largely justified, since infiltrated moisture takes considerably longer to reach a stream, and usually does not affect the storm hydrograph, which concerns a shorter time period. This is especially true in urban catchments, since in the design of the sewerage systems the engineer concerns himself with the quick and safe removal of the high storm peaks, and subsurface flow is of little consequence.

The objective of rainfall–runoff modeling is to define the magnitude and timing of excess rainfall. The latter is defined as rainfall occurring after the surface detention storage has become saturated. Mathematical determination of the excess (the transformation process that Singh and Birsoy state above) is indeed difficult. However, many models attempt to portray these processes.

INFILTRATION

> . . . infiltration theory treats the movement, in a gravitational field, of water from a surface into soil. The infiltration process involves the displacement by water of air in the voids of the soil matrix, the water moving largely in unsaturated soil (Fleming and Smiles 1975, p. 84).

Although the principles of infiltration can be explained quite simply, its mathematical definition has been the cause for much research.

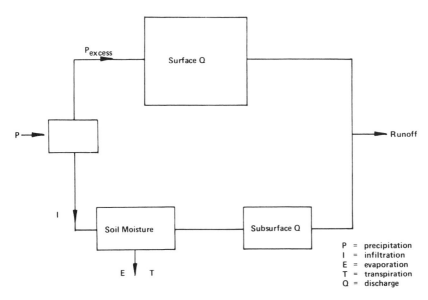

Figure 5.5. Relationship between storm rainfall and flood runoff divided into three response systems [modified from: Dooge (1968)].

Green and Ampt (1911) studied the infiltration process and presented some equations which are still commonly used today. One of their equations has been difficult to use because certain constants have to be derived by experiment. Morel-Seytoux and Khanji (1974) suggested that two constants in the Green and Ampt equation could be deduced from soil characteristics. They used one form of the Green and Ampt equation:

$$I = K(H + z_f + H_f)/z_f = A + (b/z_f)$$

where

I = infiltration rate (L/T) where L is water height and T is a unit of time
K = the fully saturated hydraulic conductivity (i.e., the conductivity of a water content equal to porosity) (L/T)
H = the height of the ponded water above the surface
z_f = the vertical extent of the saturated zone
H_f = the capillary pressure at the front expressed as a water height (L)

Horton (1940) presented a method to determine infiltration which has been widely employed in rainfall–runoff modeling. Table 5.1 lists Horton's as well as other widely used formulae for the determination of infiltration. Amorocho

Table 5.1. Approaches to the definition of infiltration.

	Horton	Kostiakov	Holtan-Overton	Philip	Natural Conditions
Formula Derived	Empirically	Empirically	Empirically	Analytically	
Soil Medium Assumed to be	Relatively Homogeneous	Relatively Homogeneous	Relatively Homogeneous	Homogeneous	Heterogeneous

Modified from: Amorocho (1967b).

(1967b) pointed out that none of these equations could be expected to clearly define infiltration over an entire basin. He based his claim on predictive uncertainty, i.e., the soils are not homogeneous over the basin and with present data collection methods (see Section 5.2), one can only estimate the soil conditions at measured points. He suggested lumping the infiltration components and using stepped inputs of rainfall (sudden applications of a constant rainfall rate). In his concluding statements, he pointed out that this method has "predictive uncertainty" problems also. Essentially, all attempts at the accurate determination of these flow directions, rates, and volumes can only be gross estimates. These are due to the nonhomogeneities of soil characteristics, soil compaction and temperature.

DEPRESSION STORAGE

If rainfall intensity exceeds the infiltration rate of the soil, depressions begin to fill. Mitchell and Jones (1976) investigated five methods of computing storage and fifteen depth-storage models, and concluded that the surface depression storage function is best described by the general model:

$$S = aD^b$$

where

S = storage in inches
D = depth above the lowest point on the surface in inches
a,b = equation parameters

More precisely, they suggested the use of a 1-inch square computational model with 0.1-inch depth increments. This method is:

$$S_r = \sum_{i=1}^{m} \sum_{j=1}^{n} (H_r - H_a)$$

where

S_r = surface storage below a reference height in cubic inches
i,j = rows and columns of point measurements, respectively
H_r = reference height in inches
H_a = point measurement of height on soil surface in inches

OVERLAND FLOW

Robertson et al. (1966) get to the root of the problem(s) related to overland flow when they write:

> ... overland flow is both unsteady and spatially varied since it is supplied by rain and depleted by infiltration, neither of which is necessarily constant with respect to time or location (p. 343).

Later in their paper, Robertson et al. attempt to remove one of the stated limitations by generation of simulated rainfall, and they still state that additional simplifying conditions were required.

Singh and Woolhiser (1976) acknowledge that surface runoff is generally recognized as a nonlinear process, and they also state that two approaches have been used by various researchers; they are the Systems Approach (Amorocho and Orlob, 1961; Bidwell, 1971) and the Hydrodynamic Approach (Chow and Ben-Zvi, 1973; Woolhiser and Liggett, 1967).

The Systems Approach develops input–output relationships without making any explicit assumptions regarding the internal structure of the system. The Hydrodynamics Approach requires the assumption that certain general laws of physics apply and further requires a geometrical abstraction of the real-world phenomenon.

Singh and Woolhiser go on in their paper to develop a hydrodynamic approach to the runoff problem. Realizing that a natural watershed has a tremendously complex surface configuration, they attempt to answer the question: "Can a simple geometry be found to adequately portray *some* important aspects of surface runoff dynamics?" Figure 5.6 is their simple representation of a natural watershed.

As the authors contend, the converging section geometry possesses some interesting features:

(1) Its discrete analog is a system composed of a cascade of unequal nonlinear reservoirs (this is a systems view).
(2) Its response is similar to that of a cascade of planes of decreasing size.
(3) The convergence may account for the concentration of runoff at the mouth of a natural watershed.

To portray the response dynamics of this watershed, they utilized kinematic wave theory, which includes the differential equation of continuity:

$$\frac{\partial h}{\partial t} + u \frac{\partial h}{\partial x} + h \frac{\partial u}{\partial x} = q(x,t) + \frac{uh}{L_o - x}$$

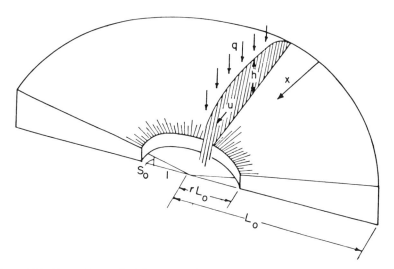

Figure 5.6. Converging overland flow model [*Source*: Singh and Buapeng, *Water Res. Bull.*, 13 (3):499–514 (1977), ©The American Water Resources Association, Minneapolis, MN].

and an approximation of the momentum equation:

$$Q = \alpha h^n \quad \text{or} \quad u = \alpha h^{n-1}$$

where

L_o = length of the section
h = the local depth of flow
u = the local mean velocity
Q = the rate of outflow per unit width
$q(x,t)$ = the rate of lateral inflow
x = the space coordinate
t = the time coordinate
n, α = the kinematic wave friction relationship parameters

Through a process called "normalization," the number of parameters are reduced to a more workable format. Normalization consists of the introduction of "normalizing" quantities, which in essence substitute for these variables and simplify the model. However, certain assumptions which may weaken the model structure are also introduced.

This surface runoff model is complex, and as the authors point out, the closer one attempts to portray real-world processes, the more complex the model becomes. As in many cases, the more complex the model, the more computer time required to solve the various components, and costs of processing increase. At some point, the researcher must reach a compromise between receiving data from the model that satisfies the requirements of the study, and yet making sure that these data may be obtained at a justifiable cost.

5.4 Urban Watershed Modeling

In the last three sections we have considered the definition and development of models and the modeling problems encountered within natural watersheds. An urban watershed not only has modified stream channels created by the increased peaks and volumes of runoff (Section 4.8), but also, in most cases, a series of totally artificial closed channels are created when sewerage is installed under a city. In the design of the latter, models are employed and, after installation, any further modeling effort must take into account the carrying capabilities of the sewerage system.

There are three basic types of sewerage systems: (1) sanitary sewerage for domestic and industrial wastes, (2) storm sewers intended to rapidly and safely convey storm runoff, and (3) combined sewerage, i.e, the first two types in one system. We will begin our discussion of urban watershed modeling with a brief presentation of sanitary sewerage design. The remainder of the section will concern discussions of storm sewerage and the widely used models for storm sewerage design—the Rational Formula and the Unit Hydrograph.

Hammer (1975) defines sewers as ". . . underground, watertight conduits for conveying waste waters by gravity flow from urban areas to points of disposal" (p. 312). Cohn (1975) wrote:

> . . . sewers are basic to community cleanliness. They guarantee that cleanliness is attainable under conditions that mass people in close proximity with their fellow men, in gregarious social structure and tight, shoulder-to-shoulder urban contacts. Sewers make possible dense urban conditions without endangering the health, comfort, safety and nicety of living of those who "density" cities. Removal of community wastewater is only part of the total urban sanitation scheme. All facets of cleanliness are involved in elimination of the offal of existence from the urban environment (p. 25).

Cohn mentions that sewer construction started as far back as 3700 B.C., stopped for various reasons during the Middle Ages (the era of plagues and pestilences) and restarted in the 18th century. Some of the very early sewerage projects, even by today's standards, are marvels of man's inventiveness. Over the past three centuries, as man has developed an increasingly better understanding of disease prevention, sewerage has become a necessary and integral part of urban structure. A major part of the urban budget is devoted to the installation and maintenance of the sewerage system. Designing such a sewer system has developed into a science, and an art at the same time. "No 'book' can be a substitute for a knowledgeable designer—a consulting engineer or a municipal design engineer."

BASIC SEWER DESIGN

Sewerage project development is divided into four phases (after the Joint Committee of the ASCE and WPCF,[43] 1969): (1) preliminary or investigative, i.e., the provision of a broad, technical and economic basis for policy decisions and final designs, (2) design, (3) construction, and (4) operation. We will briefly discuss the first two phases.

Phase 1, the preliminary or investigative phase, is extremely important in that the final products of this phase are the inputs and prime considerations of Phase 2, sewerage design. In Phase 1, growths and shifts of population are estimated, locations of industries are approximated, and in general the sewerage need requirements of a community are anticipated. In order to develop the data bases, surveys may be conducted, various population growth formulae employed, and much intuitive thinking and planning are required. Cost estimates and methods of financing are investigated, along with probable arrangement of the sewer network. The final product of this phase is mainly qualitative, and is not a detailed working design or plan from which a sewer project can be constructed. However, the final engineering report is nevertheless, as has been stated, a crucial input to the design phase. After the Joint Committee of the ASCE and WPCF, the report would include such items as:

(1) Statement of problem and review of existing conditions
(2) Capacities and conditions required to provide service for the design period
(3) Method of achieving the required service
(4) General layouts of the proposed system
(5) Establishment of applicable engineering criteria and preliminary sizing design to permit preparation of construction and operating cost estimates of sufficient accuracy to provide a firm basis for feasibility determination, financial planning, and consideration of alternative methods of solution
(6) Various available methods of financing and their applicability to the project

---

[43]ASCE is the American Society of Civil Engineers, and WPCF is the Water Pollution Control Federation.

The goal of the design phase (Phase 2) is to provide documents which form the basis for bidding and performance of the work. It begins with the layout of the sewer network and then the design flow quantities are determined (Fisher et al., 1971). In the layout, such physical factors as elevation of buildings above the sewer, types of soils and depth to bedrock, frost line depth, location of streets as well as cultural factors such as number of hookups, and type of land use, must be carefully considered.

Typical design flow criteria by land use for many cities are provided by the Joint Committee report (Tables 5.2 and 5.3). A similar summary is provided by Cohn (1975). Flow volumes may be estimated by appropriate multiplication of these values. Estimation of industrial load has recently become more complicated by the mandates of Public Law 92-500 and other environmental legislation which in many cases has caused industries to recycle water or to adopt discharge modifying measures. Cohn presents "ballpark" estimates of the anticipated discharges from a few selected industries.

A factor which may have an overriding effect on the flow in a sewer system is the Habit Curve (Cohn, 1975). For example, if the sewer flows from a residential "bedroom" community are monitored on a daily basis, it would be found that flow variations follow a typical range from a minimum during night hours to a gradual increase in the morning period, rising to the peak around suppertime. A somewhat similar curve would be observed for industries or business districts. In essence, all these land uses would place a higher demand on a sewer system at certain times during a day, and by necessity the final design must consider these varying flows.

Cohn points out two basic design principles:

(1) No sewer should be designed for flow levels more than one-half full during so-called "normal" flow periods.
(2) All sewer lines should be designed with adequate flow capacity, without surcharging (i.e., overflowing and backing up) of the conduits or without creating friction losses (damaging the sides of the conduits) which will impair the carrying capacity of the lines.

There are several formulas which engineers may employ to determine flow and the size of pipe required. All flow formulas and diagrams have in common the fact that they relate capacity, hydraulic gradients, coefficients of (inner pipe) resistance or friction, and pipe size. They differ in that a different emphasis is placed on these values (Cohn, 1975). Three of the basic formulae are the following:

(1) Manning's formula (may be used for open or closed channel flow by varying the n value):

$$V = \frac{1.486}{n} R^{2/3} S^{1/2}$$

Table 5.2. Average commercial flows.

Type of Establishment	Avg Flow (gpd/cap)
Stores, offices and small businesses	12–25
Hotels	50–150
Motels	50–125
Drive-in theaters (3 persons per car)	8–10
Schools (no showers), 8-hr period	8–35
Schools (with showers), 8-hr period	17–25
Tourist and trailer camps	80–120
Recreational and summer camps	20–25

Note: gal × 3.785 = liters.
Source: Joint Committee of the Water Pollution Control Federation and the American Society of Civil Engineers (1969), "Design and Construction of Sanitary and Storm Sewers, Water Pollution Control Federation, Manual of Practice No. 9," p. 27.

(2) Kutter formula for gravity flow conditions:

$$V = \left[\frac{\frac{1.81}{n} + 41.67 + \frac{0.0028}{s}}{1 + \frac{n}{\sqrt{R}}\left(41.67 + \frac{0.0028}{s}\right)} \right] \sqrt{RS}$$

(3) Hazen and William's formula (for closed channel conditions or sewage flows in conduits under pressure):

$$V = 1.318 C R^{0.63} S^{0.54}$$

where

V = flow velocity, in feet per second
R = hydraulic radius, in feet
S = hydraulic gradient or slope, in feet per foot
n = coefficient of roughness of pipe
C = coefficient depends on material and age of conduit

STORM SEWERAGE DESIGN

Flows at any one time in a sanitary sewer may be predicted with a great deal of confidence, because inputs are provided by variables which may be accurately monitored or computed. Storm sewerage design, however, is not so easily accomplished, because in this case the designer faces the uncertainties

Table 5.3. Some typical design flows.

City	Year and Source of Data	Average Rate of Water Consumption (gpd/cap)	Population Served (1,000s)	Average Sewage Flow (gpd/cap)	Sewer Design Basis (gpd/cap)	Remarks
Baltimore, MD	...	160	1,300	100	135 × factor	Factor 4 to 2
Berkeley, CA	...	76	113	60	92	
Boston, MA	...	145	801	140	150	Flowing half full
Cleveland, OH	1946 [6]	...	...	100	...	
Cranston, RI	1943 [6]	...	...	119	167	
Des Moines, IA	1949 [6]	...	...	100	200	
Grand Rapids, MI	...	178	200	190	200	
Greenville County, SC	1959	110	200	150	300	Service area includes City of Greenville. Sewers 24 in. and less designed to flow 1/2 full at 300 gpd/cap; sewers larger than 24 in. designed to have 1-ft freeboard
Hagerstown, MD	...	100	38	100	250	
Jefferson County, AL	...	102	500	100	300	
Johnson County, KS Mission Township Main Sewer Dist.	1958	70	70	60	1,350	Most houses have basements with exterior foundation drains
Indian Creek Main Sewer Dist.		70	30	60	675	Most houses have basements with interior foundation drains
Painesville, OH	1947 [6]	...	...	125	600	Includes infiltration and roof water

Table 5.3. (continued).

City	Year and Source of Data	Average Rate of Water Consumption (gpd/cap)	Population Served (1,000s)	Average Sewage Flow (gpd/cap)	Sewer Design Basis (gpd/cap)	Remarks
Rapid City, SD	...	122	40	121	125	New York State Board of Health Standard
Rochester, NY	1946 [6]	...	...	...	250	
Santa Monica, CA	...	137	75	92	92	
Shreveport, LA	1961	125	165	...	...	Sewer design 150 gpd/cap plus 600 gpd/acre infiltration. Sewers 24 in. and less designed to flow 1/2 full; sewers larger than 24 in. designed to have 1-ft freeboard
St. Joseph, MO	1960	...	85	125	450 / 350	Main sewers / Interceptors
Springfield, MA	1949 [5]	...	...	...	200	150 gpd/cap was used on a special project
Toledo, OH	1946 [5]	...	...	...	160	
Washington, DC Suburban Sanitary Dist.	1946 [5]	...	...	100	2 to 3.3 × avg	
Wyoming, MI	1960	150	50	82*	400	*Calculated actual domestic sewage flow not including infiltration or industrial flow

Note: gal × 3.785 = liters; gpd/acre × 0.00935 = cu m/day/ha; ft × 0.3 = m; in. × 2.54 = cm.
Modified from: Joint Committee of the Water Pollution Control Federation and the American Society of Civil Engineers (1969). "Design and Construction of Sanitary and Storm Sewers, Water Pollution Control Federation Manual of Practice No. 9," pp. 21–23.

of nonlinear responses created by rainfall distribution and surface runoff. Another factor which is not easily predicted is the impact of continually increasing impervious surface areas on runoff.

In storm sewerage design, input flows are determined by meteorological and cultural variables (i.e., the way man has modified the urban surface). Several models have been employed by engineers to determine the sizing requirements for storm sewers. The most popular is the rational method. Recently, the unit hydrograph has gained prominence. These two methods will be discussed in detail below.

The rational method was introduced by Kuichling (1889) and has become the most widely used method for estimating peak runoff rates (for areas less than five square miles) in the design of urban drainage facilities (Schaake et al., 1967), and is the following:

$$Q = CIA$$

where

Q = the design peak runoff rate in cubic feet per second
C = a dimensionless runoff coefficient
I = the average rainfall intensity (in/hr) lasting for a critical period of time, t_c
t_c = the time of concentration
A = the size of the drainage area (acres)

Viessman et al. (1977) point out:

> . . . the rationale for the method lies in the concept that application of a steady, uniform rainfall intensity will cause runoff to reach its maximum rate when all parts of the watershed are contributing to the outflow at the point of design (p. 507).

This condition is met after the elapsed time t_c which usually is taken as the time for water to flow from the most remote part of the drainage area.

Determination of the C value is difficult, because this coefficient is dependent upon many hydraulic, physiographic and climatic variables (see Section 5.2 on nonlinearities). Table 5.4 lists some values suggested by the Joint Committee Report.

I, or rainfall intensity, involves due consideration of: (1) average occurrence frequency, (2) intensity–duration characteristics of rainfall, and (3) time of concentration t_c (after Joint Committee Report). Rainfall frequency is a determining factor in sewerage design, and is the degree of protection afforded by

Table 5.4. The range of coefficients, classified with respect to the general character of the tributary area reported in use.

Description of Area	Runoff Coefficients
Business	
Downtown	0.70 to 0.95
Neighborhood	0.50 to 0.70
Residential	
Single-family	0.30 to 0.50
Multi-units, detached	0.40 to 0.60
Multi-units, attached	0.60 to 0.75
Residential (suburban)	0.25 to 0.40
Apartment	0.50 to 0.70
Industrial	
Light	0.50 to 0.80
Heavy	0.60 to 0.90
Parks, cemeteries	0.10 to 0.25
Playgrounds	0.20 to 0.35
Railroad yard	0.20 to 0.35
Unimproved	0.10 to 0.30

It often is desirable to develop a composite runoff coefficient based on the percentage of different types of surface in the drainage area. This procedure often is applied to typical "sample" blocks as a guide to selection of reasonable values of the coefficient for an entire area. Coefficients with respect to surface type currently in use are:

Character of Surface	Runoff Coefficients
Pavement	
Asphaltic and Concrete	0.70 to 0.95
Brick	0.70 to 0.85
Roofs	0.75 to 0.95
Lawns, sandy soil	
Flat, 2 percent	0.05 to 0.10
Average, 2 to 7 percent	0.10 to 0.15
Steep, 7 percent	0.15 to 0.20
Lawns, heavy soil	
Flat, 2 percent	0.13 to 0.17
Average, 2 to 7 percent	0.18 to 0.22
Steep, 7 percent	0.25 to 0.35

The coefficients in these two tabulations are applicable for storms of 5- to 10-yr frequencies. Less frequent, higher intensity storms will require the use of higher coefficients because infiltration and other losses have a proportionally smaller effect on runoff. The coefficients are based on the assumption that the design storm does not occur when the ground surface is frozen.
Source: Joint Committee of the ASCE and the WPCF (1969), "Design and Construction of Sanitary and Storm Sewers, Water Pollution Control Federation Manual of Practice No. 9," p. 51.

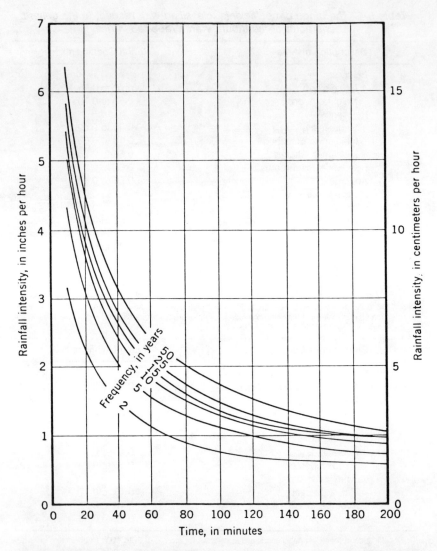

Figure 5.7. Intensity–duration rainfall curves, Boston, MA [*Source*: Joint Committee ASCE and the WPCF (1969), "Design and Construction of Sanitary and Storm Sewers," Water Pollution Control Federation Manual of Practice No. 9, p. 47].

a sewerage system. The Joint Committee Report lists several values reported by engineering offices and used in design:

(1) Storm sewers in residential areas are designed between two and fifteen years, with an average of five years.

(2) Storm sewers in commercial and high value districts, ten to fifty years, depending on economic justification

(3) Flood protection works, fifty years or more

Figure 5.7 lists intensity–duration rainfall curves for Boston, Massachusetts. A similar graph may be developed for other parts of the United States by utilizing data provided by Herschfield (1961).

An estimate of t_c (time of concentration) consists of inlet time plus the time of flow in the sewer from the most remote point in the drainage area to the design point. Surface flow is influenced by the factors outlined above under C value determination. Once storm waters gain entry into the sewers, flows may be estimated accurately by use of Manning's formula, Kutter formula or others.

Drainage area A is the one factor which can accurately be determined. This may be accomplished by examining aerial photography, surveying, or some other suitable means.

In the past several years, the rational method has been questioned. Schaake et al. (1967) tested the implicit assumption in the rational method that the frequency of occurrence of the computed design peak runoff rate is the same as the frequency of the rainfall intensity selected by the designer. They studied sewer and inlet gauge data from twenty areas in Baltimore, Maryland, ranging in size from 0.2 ac to 150 ac. The authors concluded that their results suggest that the above assumption is approximately correct. They caution, however, that this study concerned only one city, and its conclusions should not be considered universally applicable.

Since the rational formula estimates the runoff peak and not its shape (Papadakis and Preul, 1973) and does not take into account the effects of slope and watershed shape, as well as other factors, there have been attempts to modify and hopefully to improve the model (Gregory and Arnold, 1932). Nevertheless, the rational formula and its basic concepts have been and continue to be used in general engineering practice.

THE UNIT HYDROGRAPH[44]

The unit hydrograph is a "linear" model[45] (see Nash, 1959) which was developed by Sherman (1932). Gray and Wigham (1970) define the unit graph (or

---

[44]See also Section 2.1.

[45]Nonlinearities of the unit hydrograph have been subject for much research in the past twenty years (see Singh, 1964; Dooge, 1959).

the unit hydrograph) as ". . . a discharge hydrograph resulting from one inch of direct runoff generated uniformly over a tributary area at a uniform rate during a specified period of time" (p. 8.23).

Morgan and Johnson (1962) suggest that the general concept of the unit hydrograph can be summarized by the following principles:

(1) For a given drainage area, the time base of surface runoff hydrographs resulting from similar storms of equal duration are the same regardless of the intensity of rainfall.
(2) For a given drainage area, the ordinates of the surface runoff hydrographs from similar storms of equal duration are proportional to the volume of surface runoff.
(3) For a given drainage area, the time distribution of surface runoff from a particular storm period is independent of that produced by any other storm period.

Figure 5.8 demonstrates these three premises. The hydrograph in (a) was generated by 1 in. of rainfall excess falling in a unit of time; (b) shows a 2-in. rainfall having the same duration as (a). This hydrograph demonstrates principles 1 and 2 (superposition) (Machmeier and Larson, 1968), and (c) illustrates principle 3, i.e., a storm of consecutive unit time increments of 1, 2 and 0.5 in., respectively, on the drainage area.

In practice, finding a storm which displays the linear responses needed is a difficult task. Viessman et al. (1977) suggested that in deriving a unit hydrograph for a particular watershed, one must select (1) storms which occur in-

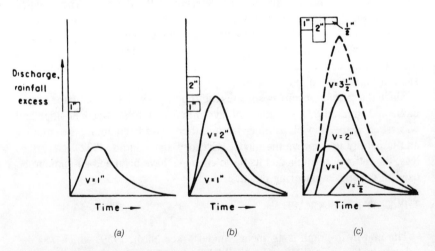

Figure 5.8. (a) Unit hydrograph; (b) unit-graph procedure applied to 2 in. of rainfall excess; (c) unit-graph procedure applied to a storm [*Source*: Morgan, P. E. and S. M. Johnson (1962), "Analysis of Synthetic Unit-Graph Methods," *J. Hyd. Div.*, ASCE, 88(HY5):201].

dividually (simple storm structure), (2) storms which have a uniform distribution of rainfall, and (3) storms which have a uniform spatial distribution. These types of meteorological phenomena place a limit on the size of the watersheds. The authors suggest an upper limit of 1000 mi², although storms occurring over 2000 mi² have been used. A lower limit of 1000 ac is suggested as a rule of thumb.

Espey, Altman and Graves (1977) point out that the best unit period is somewhat smaller than the watershed lagtime. Sherman (1949) suggested that for drainage areas in excess of 1000 mi², 12-hr units should be used in preference to 24-hr units; for areas between 100 and 1000 mi², use 6-, 8- or 12-hr units; for areas of 20 mi², use 2-hr units; for smaller areas, use a time unit which is about one-third to one-fourth of the appropriate lagtime.

After the period is chosen, the storm must produce a direct runoff of between 0.5 and 1.75 in. It is best to examine a series of suitable storms (five events) for the watershed in order to obtain an average of ordinates. The direct runoff ordinates should be reduced for each storm so that each event represents one inch of direct runoff. The final unit graph is derived by averaging ordinates of the selected events and adjusting the result to obtain 1 inch of direct runoff.

Since urban study areas usually represent a relatively small amount of area from several acres to possibly 100 mi², unit hydrographs with rather small lagtimes have been developed. Espey et al. (1977) used rainfall–runoff data from forty-one small watersheds throughout the United States to derive the 10-min unit hydrograph and its shape-forming factors [Taylor and Schwarz (1952) and Eagelson (1962) used somewhat similar methods in their analyses]. The purpose of their study was to examine a more definitive method to describe the dynamic runoff process of small watersheds. At times, the rational method may be so employed, but at other times the unit hydrograph demonstrates a greater potential.

Espey et al. found that these five physiographic characteristics are the most important indices:

A: watershed drainage area (mi²)
L: total distance (in feet) along the main channels from the point being considered to the upstream watershed divide
S: main channel slope (in feet per foot) as defined by $H/(0.8L)$, where L is the length of main channel as defined above, and H is the difference in elevation between two selected points A and B (A is at a point on the channel bottom at a distance of $0.2L$ downstream of the watershed divide, and B is a point on the stream bottom at point of consideration.)
I: extent of impervious area within the drainage area (in percent)
Φ: conveyance efficiency of the watershed's drainage system

Espey et al. noted that in an earlier study (1965) the I factor was not sufficient for describing the runoff characteristics of some urban watersheds. Therefore, a more comprehensive index was required, and Φ was introduced. Φ may be divided into Φ_1, the channel characteristics (Espey et al., 1965), i.e., the amount of channel improvement and the type of secondary drainage, and Φ_2, which accounts for flow retarding vegetation within the channel (see Table 5.5).

The empirical 10-minute hydrograph equations are shown in Table 5.6 with accompanying Figure 5.9. The authors caution that the Φ values may be difficult to derive in watersheds with atypical drainage characteristics.

The unit hydrograph, as well as the rational formula, may have in concept some basic nonlinear response problems which have been the subject for much research. However, perhaps their overall appeal and general acceptance may be attributed to the fact that they are, within reason, applicable to any urban watershed, and although they have their faults, they do describe in a broad sense the rainfall runoff response of an urban watershed.

5.5 Urban Watershed Modeling—Quantity

For some time now, the rational method and the unit hydrograph have been applied to estimate water quantity flows within the urban watershed. These procedures may be easily computed by hand. With the introduction of high speed analog and digital computers, a door was opened into the use of formerly time-consuming mathematical methods. These methods allow significantly closer approximations of the physical processes of rainfall and runoff. The simulation model was born as Fleming (1975) states:

> . . . a simulation model is the mathematical expression of the physical concepts of some phenomenon. Development of a simulation model of the hydrologic cycle involves the use of many mathematical functions to express the interrelationships between the many processes involved (p. 180).

Larson (1972) summarizes:

> . . . thanks to the development of numeric procedures for solving the partial differential equations of unsteady flow and the rapid increase in computer capabilities, these methods can now be applied to an entire watershed [sic] to generate outflow hydrographs (p. 114).

James (1965) was perhaps one of the first to see the potential capabilities of the digital computer in estimating the effects of urban development on flood peaks. Since that time, the number of urban storm water models has multiplied substantially. The basic components of most urban runoff models are shown in Figure 5.10. Models may be structured to meet particular objectives

Table 5.5. Classification of watershed drainage systems.

Φ_1	Classification
0.6	Extensive channel improvement and storm sewer system, closed conduit channel system.
0.8	Some channel improvement and storm sewers; mainly cleaning and enlargement of existing channel.
1.0	Natural channel conditions.

Φ_2	
0.0	No channel vegetation.
0.1	Light channel vegetation.
0.2	Moderate channel vegetation.
0.3	Heavy channel vegetation.

$\Phi = \Phi_1 + \Phi_2$
Source: Espey, et al. (1977), p. 10, Table 2.

Table 5.6. Ten-minute unit hydrograph equations.

Equations	Total Explained Variation
$T_R = 3.1 L^{0.23} S^{-0.25} I^{-0.18} \phi^{1.57}$	0.802
$Q = 31.62 \times 10^3 A^{0.96} T_R^{-1.07}$	0.936
$T_B = 125.89 \times 10^3 A Q^{-0.95}$	0.844
$W_{50} = 16.22 \times 10^3 A^{0.93} Q^{-0.92}$	0.943
$W_{75} = 3.24 \times 10^3 A^{0.79} Q^{-0.78}$	0.834

L = the total distance (in feet) along the main channel from the point being considered to the upstream watershed boundary
S = The main channel slope (in feet per foot) as defined by $H/(0.8L)$, where L is the main channel length as described above and H is the difference in elevation between two points, A and B. A is a point on the channel bottom at a distance of $0.2L$ downstream from the upstream watershed boundary. B is a point on the channel bottom at the downstream point being considered.
I = the impervious area within the watershed (in percent)
ϕ = the dimensionless watershed conveyance factor as described previously in the text
A = the watershed drainage area (in square miles)
T_R = the time of rise of the unit hydrograph (in minutes)
Q = the peak flow of the unit hydrograph (in cfs)
T_B = the time base of the unit hydrograph (in minutes)
W_{50} = the width of the hydrograph at 50% of the Q (in minutes)
W_{75} = the width of the unit hydrograph at 75% of Q (in minutes)
Source: Espey, Altman and Graves (1977), p. 14.

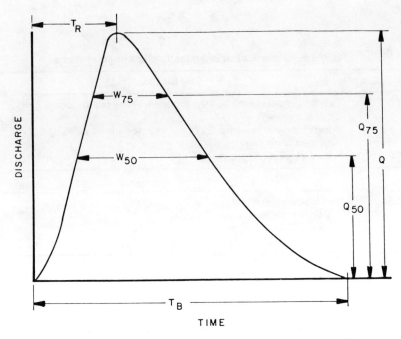

Figure 5.9. Definition of unit hydrograph parameters [*Source*: Espey et al. (1977), p. 3].

(McPherson and Schneider, 1974). In Section 5.4, we considered (1) the Rational Formula, which is used to determine the hydrograph peak, rather than shape (Papadakis and Preul, 1973), and (2) the Unit Hydrograph, which estimates peak and shape of the hydrograph. In this section, we will present several simulation models which will more comprehensively estimate the storm hydrograph arriving at the receiving water course.

The aim of any urban storm rainfall/runoff model is to reproduce as closely as possible the surface runoff hydrograph for a given storm. To achieve this end, each model must estimate a volume (x) and place this estimate in the proper time frame (y). A plot of a finite number of such points will yield a simulated storm hydrograph. Each model varies in accordance with the methods it employs in determining the various volumes and times. The volume and timing of storm surface runoff at any given time are mainly related to: (1) precipitation characteristics, (2) degree of imperviousness of the surface, (3) surface slope, and (4) watershed area. Many models attempt to define excess rainfall (that which occurs after the surface storage is saturated, or which occurs in excess of the infiltration rate of the surface). To do this, infiltration surface detention, slope and storm precipitation characteristics (i.e, duration and intensity) must be accounted for.

Volume and timing are directly related to the size of the drainage area considered. If an infinite number of highly detailed data were available on the physical characteristics of each drainage area, then these data could be used as inputs to derive very accurate estimates of the runoff process. In reality, data such as these are not readily available (also, the cost of the required computer time to adequately process such data would be prohibitive), and therefore, most models have to use a suitable subdivision (discretization) procedure. Discretization consists of subdividing the watershed into optimally discrete subwatershed areas of such a size to utilize adequately available data to produce satisfactory modeling results and stay within the computer time cost constraints.

Viessman (1966) contended that if researchers could gain a better understanding of the hydrology of individual inlet areas, since large drainage areas

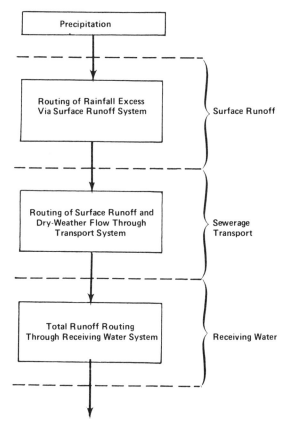

Figure 5.10. Basic components of most urban runoff models [modified from: McPherson (1978), p. 9].

are made of many small inlet areas, it would be a simple procedure to simulate the hydrograph for a large (multi-inlet) area. As he stated:

> ... this research is based on the hypothesis that for small urban areas (less than 1 acre to several acres in size) the drainage area can be considered to act as a linear reservoir. If this hypothesis can be substantiated, the promise of an improved and relatively simple procedure for predicting runoff from areas served by highly developed storm drainage systems is within sight. The hydrology of these small areas can be broken up into a subset of smaller areas, usually tributary to storm water inlets (inlet areas) (p. 406).

Viessman developed a linear reservoir model in which storage S is:

$$S = KQ \qquad (5.1)$$

where

S = storage which is directly proportional to outflow
K = (lagtime) a reservoir constant or storage coefficient having units of time (minutes, in this model)
Q = the outflow or rate of runoff

The continuity equation is:

$$I - Q = dS/dt \qquad (5.2)$$

where I = inflow to the reservoir (effective precipitation).

If we substitute the first equation with the second, and assume that at time $t = 0$ that $Q = 0$, we obtain the following equation:

$$Q = I(1 - e^{-t/k}) \qquad (5.3)$$

By assuming that effective rainfall ceases at some time after t_1 after the beginning of runoff, the outflow at any time $t > t_1$ is given by:

$$Q = Q_1 e^{-\tau/k} \qquad (5.4)$$

where $\tau = t - t_1$, or time since cessation of precipitation.

Viessman used 1-minute increments of rainfall and developed a 1-min unit hydrograph, which may be represented by Equations (5.3) and (5.4) above. Substituting 1 min for t values in (5.3) and (5.4), and replacing Q_1 in Equation (5.4) by Q_{max}, we get:

$$Q_{max} = I(1 - e^{-1/k})$$
$$Q_\tau = Q_{max} e^{-\tau/k} \qquad (5.5)$$

where $\tau = t - 1$.

These two equations are Viessman's model and other papers (Viessman, 1968; Viessman, Keating and Srinivasa, 1970) further develop and improve these concepts.

Since effective rainfall or net rainfall theoretically must equal the volume of runoff, losses must be abstracted from the total rainfall. Viessman points out that these losses are the same as encountered in natural watersheds, i.e., interception, depression storage and infiltration. However, for impervious inlet areas, the primary loss is due to depression storage, which appears to be an exponential decay-type function.

Depression storage is accounted for by deducting a specific volume from the first few minutes of rainfall. If the runoff volume still fell short of the adjusted rainfall volume, a constant rate of rainfall for the remainder of the storm was deducted. This procedure is similar to the ϕ index method used earlier by Willeke (1964). In order to effectively utilize this procedure, it is essential that: (1) for the impervious areas, some knowledge of the depression storage be available, and (2) the manner in which the antecedent conditions affect this volume be obtained. For the four areas studied by Viessman (1966), average losses ranged from a low of 0.04 in./storm to 0.14 in./storm. Willeke (1964) developed a relationship between loss per storm and mean drainage area slope for the areas studied:

$$L = 0.162 - 0.039S$$

where

L = the loss in inches per storm
S = the slope expressed in percent

The modeling procedure consists of:

(1) Abstracting these losses from the rainfall pattern
(2) Determining the value of K in Equations (5.3), (5.4) and (5.5)
(3) Substituting these values into Equations in (5.5) in the proper time frame
(4) Plotting of this point
(5) Repeating steps 1 through 4 for the next point
(6) Continuing steps 1 through 5 until the storm hydrograph is synthesized

Viessman (1966) utilized this model to derive thirty storm hydrographs. Upon comparison of these synthesized hydrographs with actual hydrographs, it was found that their peaks were in absolute error from 0.4 to 30% (some overestimated, some underestimated) with an absolute average of 9.0%. This accuracy was considered satisfactory, and although the actual storage characteristics of the impervious inlet areas were somewhat nonlinear, the author concluded that the divergence from linearity was not sufficient to warrant the more complex nonlinear analysis.

Viessman (1968) further refined the K (lagtime) value to improve the model. The average value of the optimum K ($\overline{K}$) defined by:

$$\overline{K} = 0.015(nL_2/0.015\sqrt{S})^{0.66}$$

where

n = Manning's roughness coefficient
L_2 = maximum gutter flow distance or flow length
S = mean gutter slope in feet/foot

Viessman et al. (1970) utilize the model to predict hydrographs from a (multi-inlet) 23-ac residential area, and substitute (1) a simple ϕ index, (2) an initial depression storage abstraction combined with the ϕ index, and (3) an exponential loss function (Horton type). They find that where the simple ϕ index is used, the tendency is to overpredict the rising limb of the hydrograph and underpredict the recession. The loss model under step 2 was found to improve the overall hydrograph (give the best results). The use of an exponential loss function usually fits the peaks quite well, but generally fails to reproduce faithfully the early portions of the hydrograph. In general, the authors were satisfied with results obtained from employment of this model.

During the early minutes of a storm, urban runoff is mainly derived from the impervious surfaces. In this paper, the authors attempted to define the breakpoint at which contribution from previous runoff becomes significant. They found that no valid conclusions could be drawn concerning this breakpoint.

Chen and Shubinski (1971) developed an urban storm water runoff model which will now be briefly discussed. The catchment is divided by a geometric discretization procedure into rectangular subcatchments with reasonably uniform watershed characteristics such as surface cover and ground slope. These subcatchments are not necessarily the same size; they form an aggregated system of gutters and pipes emphasizing the connectivity of flow.

Figure 5.11, center sketch, shows the accumulation of flows. Flow from subcatchment 2 flows into gutter 1, and at the corner joins with the flow from subcatchment 1, etc. The characterization of the hydraulic properties of these flows [i.e., surface area, width, ground slope, Manning's roughness coefficient, detention depth (depression storage) and infiltration rate] are the next steps in the model. Maps may be utilized to determine some of these properties, and tables and curves are used to estimate surface roughness and infiltration rates. The detention depth is taken as 0.0625 in. for impervious areas, and

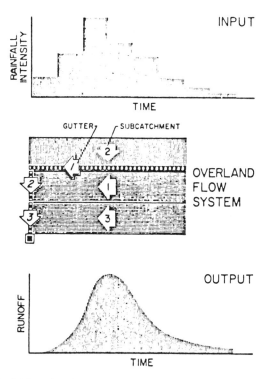

Figure 5.11. Definition of drainage system [*Source*: Chen, W. and R. P. Shubinski (1971), "Computer Simulation of Urban Storm Water Runoff," *J. Hyd. Div.*, ASCE, 97(HY2):290].

0.25 in. for pervious areas. These latter values are obtained from research completed by Crawford and Linsley (1966), and the Joint Committee of the ASCE and the WPCF (1969).

Figure 5.12 shows the flow chart of this model; an explanation of each step follows (Chen and Shubinski).

Step 1. Rainfall addition to the subcatchment is by:

$$D_1 = D_t + R_t \Delta t \ldots$$

where

D_1 = water depth after rainfall
D_t = water depth of subcatchment at time t
R_t = the intensity of rainfall in time interval Δt

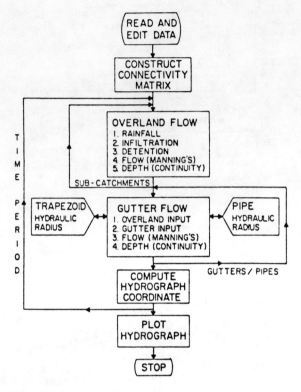

Figure 5.12. Flowchart for hydrographic computation [*Source*: Chen, W. and R. P. Shubinski (1971), "Computer Simulation of Urban Storm Water Runoff," *J. Hydr. Div.*, ASCE, 97(HY2): 292].

Step 2. Infiltration (L_t) is computed by Horton's exponential function:

$$I_t = f_o + (f_1 - f_o)e^{-\alpha t}$$

and is subtracted from water depth existing on the subcatchment:

$$D_2 = D_1 - I_t \Delta t$$

where

f_o = minimum infiltration rate as time t approaches infinity
f_1 = the maximum infiltration rate at time zero
α = the decay coefficient
D_2 = an intermediate water depth after accounting for infiltration

Step 3. If the resulting water depth of the subcatchment D_2 is larger than the specified detention depth D_d, an outflow rate is computed using Manning's equation:

$$V = \frac{1.49}{n}(D_2 - D_d)^{2/3}S^{1/2}$$

$$Q_w = VW(D_2 - D_d)$$

where

V = velocity
n = Manning's coefficient
S = the ground slope
W = width
Q_w = outflow rate

Step 4. Water depth of the subcatchment is determined by the continuity equation, resulting from the previous steps:

$$D_{t+\Delta t} = D_2 - \frac{Q_w}{A}\Delta t$$

where A = surface area of subcatchment.

Step 5. Steps 1 through 4 are repeated until computations for all subcatchments are completed.

Step 6. Inflow (Q_{in}) to a gutter is a summation of outflows from the tributary catchment ($Q_{w,i}$) and the flowrate of the immediate upstream gutters ($Q_{g,i}$);

$$Q_{in} = \Sigma Q_{w,i} + \Sigma Q_{g,i}$$

Step 7. Inflow is added to raise the existing water depth of the gutter in accordance with its geometry:

$$Y_1 = Y_t + \frac{Q_{in}}{As}\Delta t$$

where

Y_1 and Y_t = water depth of the gutter
As = the mean water surface area between Y_1 and Y_t

Step 8. Outflow is calculated for the gutter, using Manning's equation:

$$V = \frac{1.49}{n} R^{2/3} S_i^{1/2}$$

$$Q_g = VA_c$$

where

R = hydraulic radius
S_i = invert slope
A_c = cross-sectional area at Y_1

Step 9. Solution of the continuity equation to determine water depth of the gutter from the inflow and outflow. Thus,

$$Y_{t+\Delta t} = Y_1 + (Q_{in} - Q_g) \frac{\Delta t}{As}$$

Step 10. Steps 6 through 9 are repeated until all gutters are completed.
Step 11. Flows reaching the point of concern are added to produce a hydrograph coordinate along the time axis.
Step 12. Steps 1 through 11 are repeated in successive time periods until the entire hydrograph is computed.

Watt and Kidd (1975) present the Queen's University Urban Runoff Model (QUURM) and contend that this model avoids excessive complexities and introduces complexities only where justified by observations in the field. The model may be divided into two parts: (1) where the individual inlet hydrographs are determined and (2) where these hydrographs are combined and routed through the sewerage to the outlet.

In the first part, the entire catchment is subdivided into inlet subcatchments. Each subcatchment is further divided into four surface areas: (1) impervious front-paved streets, parking lots, sidewalks, and any roofs having direct access to the storm sewer; (2) pervious fronts—lawns, grass boulevards and driveways; (3) pervious and impervious backs—back lawns and roofs which drain onto the back lawns; and (4) non-contributing areas which are lower than the adjacent inlet to the storm sewerage.

For each inlet, a hydrograph is developed by use of the following models corresponding to the land phases of the hydrologic cycle.

Infiltration is developed from Horton's equation:

$$f\,\text{cap}\,(t) = f_c + (f_o - f_c)\exp(-K_f t) \tag{5.6}$$

Here $f\,\text{cap}\,(t)$, f_o and f_c are infiltration capacity at times t, 0, and ∞, respectively, and K_f is a decay constant. Equation (5.6) applies when rainfall begins at $t = 0$ and rainfall intensity $r(t)$ exceeds $f\,\text{cap}\,(t)$ for all the storm time. Where rainfall intensity is less than $f\,\text{cap}\,(t)$, the model, is modified following a suggestion after Butler (1957) to:

$$f\,\text{cap}\,(t) = f_o - K_f[F(t) - f_c t]$$

DEPRESSION STORAGE

Total volume depression storage for each area, expressed in terms of average depth, D, over this area, is assumed to be a constant. Only when rainfall rate exceeds infiltration rate are contributions made to this storage. Hence, effective rain, $i(t)$ is given by:

$$i(t) = r(t) - f(t) - d(t)$$

where

$d(t)$ = depression storage rate
$f(t)$ = infiltration rate
$r(t)$ = rainfall intensity

SURFACE FLOW

The inlet hydrograph is developed by convolution of a net rain input (1-minute intervals) with the 1-minute pulse response (somewhat like Viessman, 1966, 1968). The impulse response u is given by:

$$u(0,t) = 1/t_o \exp(-t/t_o)$$

where $t_o = nk$ (the lagtime), and the one-minute pulse response.

$$u(1,t) = \int_{t-1}^{t} u(0,\tau) \cdot d\tau$$

The ordinate (y value) of the inlet hydrograph is given by:

$$q(t) = \int_0^t i(t - \tau)u(1,\tau) \cdot d\tau$$

When A_i, the area of each type of subcatchment, is known, the total inlet hydrograph is given by:

$$Q(t) = \sum_{i=1}^{3} q_i(t) \cdot A_i$$

where A_i = the area of type i in the subcatchment.

SEWER ROUTING

The authors employ the time offset method in the sewer routing procedure. Beginning at the remote inlets where the complete storm hydrographs are known, a representative discharge is evaluated. This flow is defined as the average discharge occurring over the central 50% of this hydrograph. Then, using Manning's equation, a representative velocity is computed for the pipe immediately downstream of the inlet. The pipe slope is used as the slope term, and the published n value for the pipe is the n term. The length of the pipe divided by the representative velocity supplies the time offset.

Watt and Kidd applied the QUURM to ten rainfall events occurring in the summers of 1972 and 1973 over a low density residential housing area in Kingston, Ontario. Area types (as previously described) were 1, 2, and 3, with areas of 25.1, 27.7 and 34.2 ac, respectively. For purposes of comparison, the EPA Storm Water Management Model (SWMM) was also used in the simulation. Figure 5.13 shows the results from one storm. The authors pointed out that their purpose was not to compare in detail QUURM and SWMM. Nevertheless, QUURM simulated the observed storm hydrographs at least as well as the calibrated SWMM.

5.6 Urban Watershed Modeling—Quality

ACTUAL MODELS

Viessman et al. (1977) write:

> ... a water quality model is a mathematical statement or set of statements that equate water quality at a point of interest to causative factors. In general, water

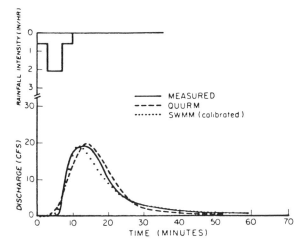

Figure 5.13. Observed and simulated hydrographs for storm of 8/23/72 on Calvin Park catchment [*Source*: Watt, W. E. and C. H. R. Kidd (1975), "QUURM—A Realistic Urban Runoff Model," *J. Hydr.*, 27:233].

quality models are designed to (1) accept as input, constituent concentration versus time at points of entry to the system; (2) simulate the mixing and reaction kinetics of the system; and (3) synthesize a time-distributed output at the system outlet (p. 647).

These three points essentially outline the major problems to be considered in urban water quality modeling. Water quantity has a direct bearing on point 1, and a significant influence on point 2.[46]

The roughness, imperviousness, slope and temperature of the urban surface at any given time has a direct bearing on the rate and volume of storm runoff. The capabilities of these runoff waters to dislodge and carry urban dust and refuse are directly related to the rate and volume of runoff. In a similar manner, the interactions between the organic and inorganic portions of the refuse with the runoff waters and each other are also influenced by the rate and the volume of the runoff waters.

All these factors may be quite easy to describe qualitatively and conceptually.[47] However, their quantitative prediction and definition has occupied much research and computer time. In this section, in discussion of Viessman's three points above, we will: (1) present one approach to water quality modeling of urban surface runoff, (2) discuss the reaction kinetics in a river and present modeling efforts, and (3) present a conceptual model of the complete process.

[46]See Pitt and Field (1977) for more discussion on the quality–quantity relationships.
[47]See McElroy et al. (1976) for an excellent discussion on these factors.

MODELING OF WATER QUALITY IN SURFACE RUNOFF

The MUNP (Management of Urban Nonpoint Pollution) Model is outlined by Sutherland and McCuen (1978), and the model flow chart is shown in Figure 5.14. As the authors point out, this model is comparable in structure to the Storage, Treatment, Overflow, Runoff Model (STORM) of the Hydrologic Engineering Center, U.S. Army Corps of Engineers (1976).

The accumulation component computes an estimate of the pollutants which have accumulated on the street surfaces. This portion of the model is based on work by Sartor and Boyd (1972). This component is a function of land use (either residential, commercial or industrial), pavement type and condition, traffic volume, population density and the length of time since street sweeping or rainfall. Equations are developed for each land use incorporating these variables (Table 5.7). Next, it was necessary to quantify various qualitative variables such as heavy vs. light traffic volumes, and fair vs. good pavement conditions. Final accumulation equations were derived for single-family detached housing and multi-family housing, and commercial and industrial land uses were developed as a function of pavement condition and traffic volume.

The rainfall component is developed also after research by Sartor and Boyd (1972) in addition to an equation by Yalin (1963) and modified by Foster and Meyer (1972). Yalin assumed that sediment motion would begin when the critical lift force was exceeded. Sartor and Boyd determined that about 88% of the accumulated total solids loadings were located within 12 in. of the curb. Using these methods, Sutherland and McCuen simplified the modeling of pollutant removal to include this area, and assumed that material transport was due to shallow, open channel gutter flow. To predict accurately the radius of the gutter flow, the Basic Inlet Hydrograph Model (BIHM) as developed by Ragan and Root (1974) was used in conjunction with the modified Yalin equation.

In considering the rate of tranport of soil particles, particle size is one of the most important variables. Six particle size ranges were defined by Sartor and Boyd and used in the model. Another factor to be considered is the amount of

Table 5.7. Preliminary equations for accumulation component.

Land Use	Correlation Coefficient	Standard Error (lbs/curb mile)	Equations
Industrial	0.91	194	$P_I = 1388(1 - e^{-0.19t})$
Commercial	0.71	165	$P_C = 500(1 - e^{-0.335t})$
Residential	0.49	268	$P_R = 1089t/(1.0 + 1.3t)$

t = time, in days, since rainfall or street-sweeping
P = accumulation of total solids in pounds/curb mile
Source: Sutherland and McCuen (1978), p. 412, *Water Resources Bull.*, 14(2):409–428.
©The American Water Resources Association, Minneapolis, MN.

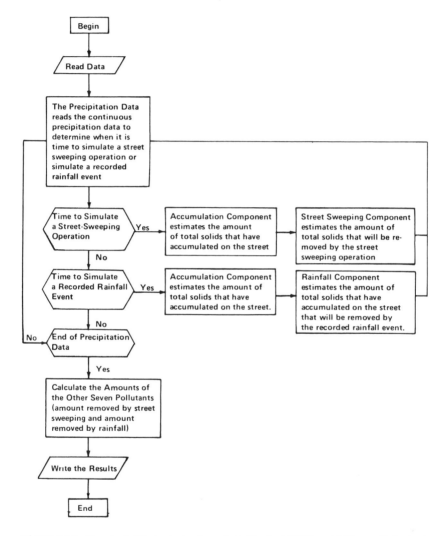

Figure 5.14. Flowchart of the basic structure adopted for the MUNP model [*Source*: Sutherland and McCuen (1978), p. 411, *Water Resources Bull.*, 14(2):409–428. ©The American Water Resources Association, Minneapolis, MN].

material initially available for transport in each reach of the gutter. When surface runoff achieves a required rate, it will begin to carry off certain particle sizes. These sizes are automatically eliminated from further consideration, and the runoff gains an enhanced competence to carry off the remaining material. The competence of runoff waters is obviously also related to such factors as slope and imperviousness.

After the initial total solids loading is received from the accumulation component, the rainfall component estimates the removal percentages in each of the six particle size ranges that can be expected from a 0.5-in. total volume rainfall. If the actual rainfall is not 0.5 in., then the value of K_j is adjusted as in the following equation:

$$TS_j = K_j(TS_i)$$

where

TS_j = the percentage removal of total solids in a particle size range due to a total rainfall volume j (in.)
TS_i = the percentage removal of total solids in a particular size range due to a total rainfall volume of 0.5 in.
K_j = a factor that relates TS_j to TS_i

An estimate of the removal of the accumulated pollutants by street sweepers is obtained from the street-sweeping component. Street-sweeping effectiveness is related to (1) the particle size range of the pollutants, (2) type of sweeper (motorized or vacuum), (3) forward speed of the street sweeper, and (4) pavement type. This component operates independently of the accumulation component. The street-sweeping component is based on research by Lee et al. (1959) and Clark et al. (1963).

In their abstract for this paper, Sutherland and McCuen contend that the MUNP model can simulate the following pollutants: total solids or sediment-like material, volatile solids, five-day biochemical oxygen demand, chemical oxygen demand, Kjeldahl nitrogen, nitrates, phosphates and total heavy metals. In the paper, the former is simulated to estimate the efficiency of street sweepers in the Washington, DC area. Figures 5.15 and 5.16 show the results of the simulations for four hypothetical locations. The frequency of street sweeping has the greatest effect on the removal efficiency. Vacuumized sweepers seem to be about 5% more efficient than the motorized ones. Street sweepers operating at 3 mph are about 3% more efficient than ones operating at 4 mph. The most efficient street-sweeping operation for the four areas was accomplished by a 3-mph vacuumized sweeper at an interval of 2 days.

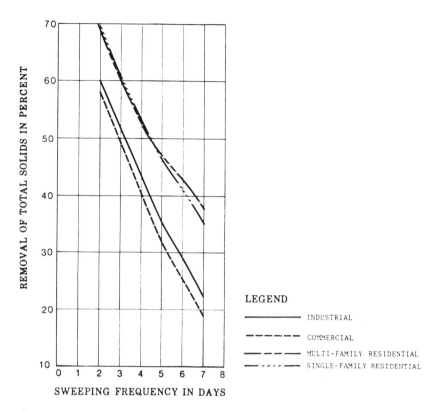

Figure 5.15. Effect of land use of the removal of total solids by street sweeping: vacuumized sweeper operating at 3 mph [*Source*: Sutherland and McCuen (1978), p. 421, *Water Resources Bull.*, 14(2):409–428, ©The American Water Resources Association, Minneapolis, MN].

STREAM WATER QUALITY MODELING

After the water quantity factors of volume and rate have affected the surface runoff (inorganic–organic) water quality factors as outlined above, this water mixture enters the stream system and may, in a similar manner, bring about similar water quantity–water quality interactions. Viessman (1969) points out:

> ... a short, high-peaked surface runoff hydrograph of suspended matter could be expected to affect more seriously a receiving water than a hydrograph which released the same volume of suspended matter over an extended period (p. 90).

He states later in the paper that some studies have suggested that chemical quality data for a stream may be related to discharge by:

$$C = {}_kQ^n \quad [\text{sic}, kQ^n] \tag{5.7}$$

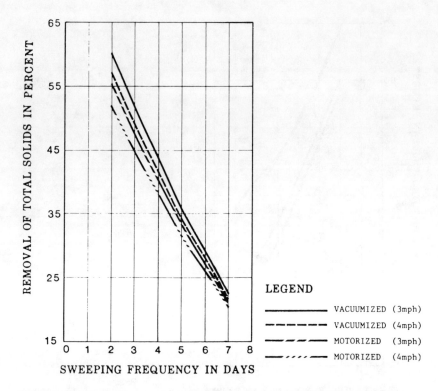

Figure 5.16. Effects of sweeper characteristics on the removal of total solids: industrial land use [*Source*: Sutherland and McCuen (1978), p. 422, *Water Resources Bull.*, 14(2):409–428, ©The American Water Resources Association, Minneapolis, MN].

where

C = the constituent concentration
Q = streamflow
k,n = regression parameters

Ledbetter and Gloyna (1964) developed scatter diagrams for this equation where C equalled mineral concentrations (chlorides). They suggested that a considerable improvement in prediction could be made by considering b as a variable instead of a constant in the equation:

$$b = pQ^n \qquad (5.8)$$

where

b = the exponent n in Equation (5.7)
p,n = regression coefficients

Ledbetter and Gloyna go on to conclude that:

> ... the inorganic quality of streamflows may be estimated from the quantity of flow by a hyperbolic form, a logarithmic form with constant exponent, or a logarithmic form with a variable exponent relationship (p. 150).

The above equations present a sound case for the prediction of the dilution capabilities of a stream. Let us now investigate more closely the actual mixing phenomena occurring in rivers.

The simulation of inorganic flows is one part of the problem in stream water quality modeling. The other major part is the simulation of the organic portion. Inorganic constituents are called "conservative" in that they do not decay appreciably over time. On the other hand, oxygen-demanding organic constituents are classified as nonconservative constituents, since they do decay with time.[48]

As assessment of the nonconservative constituents at any one time obviously is complicated by the time-varying nature of the pollutant.

Hays and Krenkel (1968) theoretically discussed the mixing phenomena in rivers. They present the continuity equation for the transport of a conservative tracer in a river by the following differential equation:

$$\frac{\partial c}{\partial t} = \frac{\partial}{\partial x}\left(D^{mt}\frac{\partial c}{\partial x}\right) + \frac{\partial}{\partial y}\left(D^{mt}\frac{\partial c}{\partial y}\right) + \frac{\partial}{\partial z}\left(D^{mt}\frac{\partial c}{\partial z}\right)$$
$$- \frac{\partial(v_x c)}{\partial x} - \frac{\partial(v_y c)}{\partial y} - \frac{\partial(v_z c)}{\partial z}$$

where

c = tracer concentration
x, y, z = coordinate directions with the corresponding velocities v_x, v_y, v_z
D^{mt} = includes the effects of both molecular and turbulent diffusivity and may be a function of x, y, and z (p. 111)

This equation, needless to say, is formidable in application, as the authors admit, and, as is the case with many other formulae in engineering and the physical sciences, certain assumptions must be made in order to reduce the particular formula to a more useable and applicable form. The authors point out that earlier investigators selected the simple geometry of the pipe on which to focus their efforts, and developed the following transport equation:

$$\frac{\partial c}{\partial t} = \frac{1}{r}\frac{\partial}{\partial r}\, rD^{mt}\frac{\partial c}{\partial r} + D^{mt}\frac{\partial^2 c}{\partial x^2} - v_x\frac{\partial c}{\partial x}$$

---

[48]See Viessman et al. (1977) for a brief discussion of conservative and nonconservative constituents.

where r = the radial coordinate. The restrictions on this equation (the compromises which must be accepted to provide a more easily useable formula) are: (1) zero velocity in the radial and angular directions and (2) no concentration gradients in the angular direction.

One researcher, Taylor (1954), by considering only the cross-sectional average concentration and using empirical expressions for velocity and the Reynolds analogy to define the variation of v_x and D^m and r, further simplified the latter formula. This formula is a one-dimensional dispersed flow model:

$$\frac{\partial c}{\partial t} = D_L \frac{\partial^2 c}{\partial x^2} - \bar{u} \frac{\partial c}{\partial x}$$

where

D_L = the longitudinal mixing coefficient
$\bar{u}$ = the average velocity (Hays and Krenkel)

This model and its variations have been used by numerous researchers in the development of river water quality models.[49]

Another model which has a fundamental basis and is used widely in concept is the conservation of mass and momentum or the following (O'Loughlin, 1975):

$$\frac{\partial v_i}{\partial t} = (I - O)_i + S_i$$

where

v_i = the index variable, the concentrations of the substance of interest
$(I - O)_i$ = the difference between rates of mass input and output including advection and dispersion
S_i = the net rate of mass addition from internal sources or sinks

Thus far, we have investigated the concepts on which many models are based. Now an actual stream model will be outlined.

Willis, Anderson and Dracup (1975) developed the EDI-Qual-I model which they used in the study of the Truckee-Carson River Basin in California and Nevada. The authors point out that the mass transport of constituents is accomplished by advection and dispersion in a river system, and use the following formula to express these concepts:

$$A \frac{\partial c}{\partial t} = \frac{\partial \left(A D_L \frac{\partial c}{\partial x} \right)}{\partial x} - \frac{\partial (A \bar{u} c)}{\partial x} \pm AS$$

---

[49] Yotsukura and Sayre (1976) propose the use of a two-dimensional mixing formulation.

where

D_L = longitudinal dispersion coefficient
$\bar{u}$ = mean stream velocity
A = cross-sectional area
c = concentration of constituent under consideration
t = some point in time
x = some point along the longitudinal axis of the stream (x-axis)
S = sources or sinks of nonconservative constituent

The modeling procedure consists of (1) breaking up the river system into reaches (each reach has essentially constant hydrologic and quality conditions), (2) routing the governing equations over each reach, and (3) determining initial concentrations of conservative and nonconservative constituents for each reach.

Both conservative (minerals) and nonconservative substances [i.e., DO, BOD, phosphorous, coliforms (total and fecal), ammonia, nitrites, nitrates, and chlorophyll-a] may be simulated by the EDI-QUAL-I Model. The oxygen budget considers the effects of nitrification (ammonia reduced to nitrite and continued reduction to nitrate) as well as the oxygen supplied by algae (phytoplankton and benthic plankton). Elevation and temperature data are considered on a reach-by-reach basis.

The authors write that the model performed well; however, they caution that:

> . . . the model is only as good as the input data and the reader is advised to review carefully the data contained in the input files of the appropriate verification run before drawing any conclusions from the simulation curves (p. 253).

This consideration is, of course, true with any model. The system equations for this model cover approximately two pages of text, and will not be presented here.

Jewell and Adrian (1982) discuss the statistical selection of significant variables in storm water quality models. They write:

> Two phenomena make derivation of functional relationships between storm water pollutant loadings and various independent variables difficult. First is the inherent variability of storm water data caused by the random component of storm events and sampling and analysis errors. Second is the large number of independent variables and parameters that may influence storm water pollution washoff (p. 489).

The authors present the results of the application of fourteen different models on data from 261 storm events from twenty-six basins in twelve geographical areas. They conclude:

> . . . [N]o one model is consistently better than others in predicting stormwater pollutant washoff for several basins from different geographical areas (p. 499).

CONCEPTUAL MODEL OF URBAN WATER QUALITY MODELING

McPherson (1978) divided urban runoff models into three components: surface runoff, sewerage transport and receiving water (Figure 5.17). He writes that this figure depicts ". . . the principal functional components of the most comprehensive urban runoff models." (p. 8). He points out further:

> . . . whether areawide urban runoff planning should involve the use of such comprehensive models is definitely a matter of opinion and is controversial. However, quantity and quality accounting of "surface runoff" and "receiving water" are necessary elements of any areawide appraisal, and the degree of detail that should be incorporated requires careful weighing of a number of complex considerations (p. 8).

These sentences form a proper conclusion to this section.

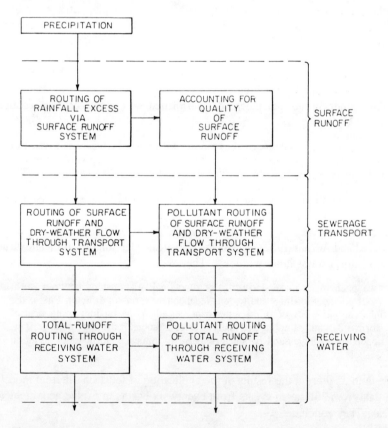

Figure 5.17. Components of urban runoff models [*Source*: McPherson (1978), p. 9].

5.7 Summary

A model is defined after Snyder and Stall (1965) as previously stated:

> ... simply the symbolic form in which a physical principle is expressed. It is an equation or formula, but with the extremely important distinction that it was built by consideration of the pertinent physical principles, operated by logic, and modified by experimental judgement and plain intuition (p. 87).

Mathematical model building consists of applying mathematics to a real-life situation. This is demonstrated through the development of a hypothetical linear model. The example of the real-world situation is a 100-gal drum located under a spout draining the roof of a small shed. By noting precipitation data from a nearby rain gauge, measuring volumetric changes in the barrel after each rainstorm and plotting these values, a mathematical relationship is developed.

It is determined that after a 1-in. rainfall, the amount of water in the drum is increased by 10 gal. Data derived from the model are found to track the curve very well (display linear characteristics) with the exception of rainfalls less than 1 in., which are nonlinear in this range. Through the use of a curve-fitting procedure (optimization) these lower points are brought up closer to the curve. This model is called a simple (parametric) rainfall/runoff (volume) regression model. If a more intensive effort were made to mathematically portray the different components of this rainfall/runoff process, the new model developed could be called "deterministic." A stochastic model has probability associated with output.

The modeling of a natural watershed is complicated by the inherent nonlinearity of its processes. Amorocho (1967a) classified nonlinear problems as involving: (1) time variability of watersheds, (2) uncertainty with respect to the time and space distribution of the inputs of the hydrologic system, and (3) inherent nonlinearity of the processes of mass and energy transfer that constitute the hydrologic cycle. The closer the researcher tries to simulate the watershed response, the more unwieldy the model becomes. The result of these nonlinear response characteristics is to cause the model builder to make assumptions and compromises. Several modeling approaches are presented; each approach has its advantages and disadvantages when examined closely.

Rainfall/runoff modeling is a problem which has occupied much time and effort on the behalf of many hydrologists. It is one of the fundamentals of hydrology which remains adamant in its capability not to be mastered or understood. The objective of rainfall/runoff modeling is to define the magnitude and timing of excess rainfall (i.e., rainfall occurring after the surface detention storage has become saturated). Several approaches to infiltration, depression storage and overland flow are presented.

Sewers are part of the hydrologic environment of an urban area. There are three basic types of sewers, sanitary, storm and combined. Basic sanitary sewer design is divided into four phases following the Joint Committee of the ASCE and WPCF (1969) report. The first two are concerned with planning and design, and the remainder with construction and operation.

Storm sewerage design has been accomplished by use of the rational formula and the unit hydrograph has gained some prominence. The rational method is used to estimate the peak runoff rates, usually for areas of less than five square miles. Viessman et al. (1977) explain:

> . . . the rationale for the method lies in the concept that application of a steady uniform rainfall intensity will cause runoff to reach its maximum rate when all parts of the watershed are contributing to the outflow at the point of design (p. 507).

The rational method, even though questioned by several authors, has been found to be approximately correct. Perhaps the major reasons for its widespread acceptance and use are that its computation is easy, and by experience and tradition it has proven itself useful.

The unit hydrograph not only estimates peak flow, but also the shape of the hydrograph. The unit hydrograph is a hydrograph resulting from 1 in. of direct runoff generated uniformly over a tributary area at a uniform rate during a specified period of time (Gray and Wigham, 1970). For the watershed under study, storms are selected meeting these criteria listed above, and using these events a researcher can manipulate the data to arrive at a unit hydrograph by considering the effects of Φ_1 channel characteristics, and Φ_2 flow-retarding vegetation in the channel.

With the introduction of high-speed analog and digital computers, formerly time-consuming mathematical methods which were thought too cumbersome to utilize could now be attempted. The simulation model was born, and a threshold was crossed into the development of models which would yield results far superior to those of the rational formula and the unit hydrograph. Viessman (1966, 1968) and Viessman et al. (1970) developed models on the concept that an urban watershed is made up of a finite number of small catchments, and if one can accurately determine what is occurring in these small areas, then one can combine all these individual responses and estimate what the hydrograph should be for the entire watershed. Viessman considers these small areas as acting as linear reservoirs. After comparing model hydrographs with the actual hydrographs, discrepancies were found; however, Viessman was satisfied with his overall results. Urban watershed models by Chen and Shubinski (1971) and Watt and Kidd (1975) are presented. These models attempt to portray comprehensively the runoff process, and yield good results when pertinent input data are available and when calibrated properly.

Urban water quality modeling is divided into the land phase and the in-stream phase. The MUNP (Management of Urban Nonpoint Pollution) Model as outlined by Sutherland and McCuen (1978) describes very well the various surface processes and their effect on water quality. The model was developed using the research of others such as Sarton and Boyd (1972), Yalin (1963), Foster and Meyer (1972), Ragan and Root (1974), Lee et al. (1959) and Clark et al. (1963). The MUNP Model consists of an accumulation component, a rainfall component, and a street-sweeping component. Each component is logically developed, and the interrelationships between components are accounted for.

Instream water quality modeling is divided into inorganic and organic simulation. Ledbetter and Gloyna (1964) developed a regression equation for mineral concentrations (chlorides) and their dilution in a stream. They concluded that:

> ... the inorganic quality of streamflows may be estimated from the quantity of flow by a hyperbolic form, a logarithmic form, with constant exponent, or a logarithmic form with a variable exponent relationship (p. 150).

Organic dispersal simulation is a formidable problem due to the fact that these constituents decay with time (they are nonconservative). Hays and Krenkel (1968) present some differential equations to express the mixing phenomena of nonconservatives. These equations are too unwieldy to attempt solution, and assumptions are made to reduce them to a more applicable workable format.

The EDI-QUAL-I Model developed by Willis, Anderson and Dracup (1975) is described. The modeling procedure consists of (1) breaking up the river system into reaches, (2) routing the governing equations over each reach, and (3) determining initial concentrations of conservative and nonconservative constituents for each reach. An important point brought out by the authors will be repeated here:

> ... the model is only as good as the input data, and the reader is advised to review carefully the data contained in the input files (p. 253).

McPherson (1978) divides urban runoff models into three components: surface runoff, sewerage transport and receiving water. These components provide conceptual divisions for any urban runoff model.

5.8 Exercises

Considering an urban stream with which you are very familiar:

1 After examining the available data, which modeling approach (i.e., deterministic, parametric or stochastic) do you think would yield the most reliable results if you were to consider: 1) changes in surface runoff, 2) peak flows, 3) salt concentrations in surface runoff, and 4) lead concentrations in surface runoff?

a) Explain the difference(s) between modeling for surface runoff and peak flows. Do you think they are significant?

b) Explain the difference(s) between modeling for salt concentrations and lead concentrations in surface runoff. Do you think they are significant?

2 Plot a unit hydrograph (as shown in 5.8) for a particular point along the urban stream for a 1" storm, a 2" storm, and a 3½" storm.

a) Do the shapes of the hydrographs differ? Why is this so?

3 If you were to redesign the sewerage system for an urban area with which you are very familiar, which approach would you use? Defend your choice.

References

AMOROCHO, J. and G. T. Orlob. "Nonlinear Analysis of Hydrologic Systems," University of California Water Resources Center Contribution No. 40, 147 pp. (1961).

AMOROCHO, J. "Measures of the Linearity of Hydrologic Systems," *J. Geophys. Res.*, 68(8): 2237–2249 (1963).

AMOROCHO, J. "Predicting Storm Runoff on Small Experimental Watersheds," *J. Hyd. Div.*, ASCE, (HY2):185–191 (1961).

AMOROCHO, J. "The Nonlinear Prediction Problem in the Study of the Runoff Cycle," *Water Resources Res.*, 3(3):861–880 (1967a).

AMOROCHO, J. "Role of Infiltration in Nonlinear Watershed Analysis Processes," *Trans. Am. Soc. Agric. Eng.*, pp. 398–404 (1967b).

BIDWELL, V. J. "Regression Analysis of Nonlinear Catchment Systems," *Water Resources Res.*, 7(5):1118–1126 (1971).

CHAPMAN, T. G. and F. X. Dunin, eds. *Prediction in Catchment Hydrology*. Australian Academy of Science. Netly, S. Australia:The Griffin Press, 498 pp. (1975).

CHEN, W. and R. P. Shubinski. "Computer Simulation of Urban Storm Water Runoff," *J. Hyd. Div.*, ASCE, 97(HY2):289–302 (1971).

CHIU, C.-L. and R. P. Bittler. "Linear Time-Varying Model of Rainfall–Runoff Relation," *Water Resources Res.*, 5(2):426–437 (1969).

CHIU, C.-L. and J. T. Huang. "Nonlinear Time-Varying Model of Rainfall–Runoff Relation," *Water Resources Res.*, 6(5):1277–1286 (1970).

CHORLEY, R. J. and B. A. Kennedy. *Physical Geography: A Systems Approach*. London: Prentice-Hall Inc., 370 pp. (1971).

CHOW, V. T. and A. Ben-Zvi. "Hydrodynamic Modeling of Two-Dimensional Watershed Flow," *J. Hyd. Div.*, ASCE, 99(11):2023–2040 (1973).

CLARK, D. E., JR. and W. C. Cobbin. "Removal Effectiveness of Simulated Dry Fallout from Paved Areas by Motorized and Vacuumized Street Sweepers," U.S. Naval Radiological Defense Laboratory, Alexandria, Va., USNRDL-TR-746, 100 pp. (1963).

CLARKE, R. T. "A Review of Some Mathematical Models used in Hydrology, with Observations on Their Calibration and Use," *J. Hydrology*, 19(1):1–20 (1973).

COHN, M. M. "By the Magic of Chemistry: Pipelines for Progress," Report by Certain-teed Products Corporation, Valley Forge, PA, 241 pp. (1975).

COSGRIFF, R. L. *Nonlinear Control Systems.* New York:McGraw-Hill Book Company, 328 pp. (1958).

CRAWFORD, N. H. and R. K. Linsley. "The Stanford Watershed Model MK IV," Technical Report No. 39, Department of Civil Engineering, Stanford University, Menlo Park, CA (1966).

DAWDY, D. R. "Considerations Involved in Evaluating Mathematical Modeling of Urban Hydrologic Systems," U.S. Geological Survey Water Supply Paper No. 1591-D, Washington, DC (1969).

DOOGE, J. C. I. "A General Theory of the Unit Hydrograph," *J. Geophys. Res.*, 64(2):241–256 (1959).

DOOGE, J. C. I. "The Hydrologic Cycle as a Closed System," *Bull. Int. Assoc. Sci. Hydrol.*, 13(1): 58–68 (1968).

DUNIN, F. X. "Use of Physical Process Models," in *Prediction in Catchment Hydrology.* T. G. Chapman and F. X. Dunin, eds. Netly, S. Australia:The Griffin Press, pp. 277–291 (1975).

EAGELSON, P. S. "Unit Hydrograph Characteristics for Sewered Areas," *J. Hyd. Div.*, ASCE, 88 (HY2):1–25 (1962).

ESPEY, W. H., JR. et al. "A Study of Some Effects of Urbanization on Storm Runoff from a Small Watershed," Technical Report HYD 07-6501, CRWR-2, Center for Research in Water Resources, Department of Civil Engineering, University of Texas, Austin, Texas (1965).

ESPEY, W. H. JR. et al. "Nomographs for Ten-Minute Unit Hydrographs for Small Urban Watersheds," ASCE Urban Water Resources Research Program, Technical Memorandum No. 32 (December, 1977).

FISHER, J. M. et al. "Design of Sewer Systems" *Water Resources Bull.*, 7(2):294–302 (1971).

FLEMING, G. *Computer Simulation Techniques in Hydrology.* New York:Elsevier Publishing Co., 333 pp. (1975).

FLEMING, P. M. and D. E. Smiles. "Infiltration of Water into Soil," in *Prediction in Catchment Hydrology.* T. G. Chapman and F. X. Dunin, eds. Netly, S. Australia:The Griffin Press, pp. 83–110 (1975).

FOSTER, G. R. and L. D. Meyer. "Transport of Soil Particles by Shallow Flow," *Trans. Am. Soc. Agric. Eng.*, 15(1):99–102 (1972).

GRAY, D. M. *Handbook on the Principles of Hydrology.* Ottawa, Canada:Canadian National Committee for the International Hydrological Decade, 1500 pp. (1970).

GRAY, D. M. and J. M. Wigham. "Peak Flow-Rainfall Events," in *Handbook on the Principles of Hydrology.* D. M. Gray, ed. Ottawa, Canada:Canadian National Committee for the International Hydrological Decade, Section VIII (1970).

GREEN, W. H. and C. A. Ampt. "Studies on Soil Physics, 1. Flow of Air and Water through Soils," *J. Agr. Sci.*, 4:1–24 (1911).

GREGORY, R. L. and C. E. Arnold. "Runoff-Rational Formulas," *Trans. ASCE*, 96:1038–1099 (1932).

GUPTA, A. and H. Fox. "Effects of High-Magnitude Floods on Channel Form: A Case Study in the Maryland Piedmont," *Water Resources Res.*, 10(3):499–509 (1974).

HAMMER, M. J. *Water and Wastewater Technology.* New York:John Wiley and Sons, Inc., 502 pp. (1975).

HAYS, J. R. and P. A. Krenkel. "Mathematical Modeling of Mixing Phenomena in Rivers," in *Advances in Water Quality Improvement.* E. F. Gloyna and W. W. Eckenfelder, eds. Austin: University of Texas Press, pp. 111–123 (1968).

HERSHFIELD, D. M. "Rainfall Frequency Atlas of the United States," U.S. Weather Bureau, Technical Paper No. 40 (1961).

HORTON, R. E. "An Approach Toward the Physical Interpretation of the Infiltration Equation," *Soil Sci. Soc. Proc.*, 5:399–417 (1940).

HUGGINS, L. F. and E. J. Monke. "A Mathematical Model for Simulating the Hydrologic Response of a Watershed," *Water Resources Res.*, 4(3):529–539 (1968).

HYDROLOGIC ENGINEERING CENTER, CORPS OF ENGINEERS. "Urban Storm Water Runoff," STORM Computer Program No. 723-S8-L2520 (1975).

JACOBY, S. L. S. "A Mathematical Model for Nonlinear Hydrologic Systems," *J. Geophysical Res.*, 71(20):4811–4824 (1966).

JAMES, L. D. "Using a Digital Computer to Estimate the Effects of Urban Development on Flood Peaks," *Water Resources Res.*, 1(2):223–234 (1965).

JEWELL, T. K. and D. D. Adrian. "Statistical Analysis to Derive Improved Stormwater Quality Models," *Journal of the Water Pollution Control Federation*, 54(5):489–499 (1982).

JOINT COMMITTEE OF THE AMERICAN SOCIETY OF CIVIL ENGINEERS AND THE WATER POLLUTION CONTROL FEDERATION. "Design and Construction of Sanitary and Storm Sewers," WPCF Manual of Practice No. 9, Washington, DC, pp. 2–3 (1969).

KUICHLING, E. "The Relation Between the Rainfall and the Discharge of Sewers in Populous Districts," *Trans. ASCE*, 20:1–60 (1889).

LARSON, C. L. "Using Hydrologic Models to Predict the Effects of Watershed Modification," in *Watersheds in Transition*. by S. C. Csallany et al., eds. Urbana, Illinois: American Water Resources Association, pp. 113–117 (1972).

LEDBETTER, J. O. and E. F. Gloyna. "Predictive Techniques for Water Quality Inorganics," *J. San. Eng. Div.*, ASCE, 90(SA1)I:127–151 (1964).

LEE, H. et al. "Stoneman II Tests of Reclamation Performance Characteristics of Dry Decontamination Procedures," U.S. Naval Radiological Defense Laboratory, Alexandria, VA, USNRDL-TR-336 (1959).

MCELROY, A. D. et al. "Loading Functions for Assessment of Water Pollution from Nonpoint Sources," EPA-600/2-76-151, 445 pp. (May, 1976).

MCLONE, R. R. "Mathematical Modeling: The Art of Applying Mathematics," in *Mathematical Modeling*. J. G. Andrews and R. R. McLone, eds. England: Butterworths Publishing Co., Chapter 1 (1976).

MCPHERSON, M. B. and W. J. Schneider. "Problems in Modeling Urban Watersheds," *Water Resources Res.*, 10(3):434–440 (1974).

MCPHERSON, M. B. "Urban Runoff Control, Quantity and Quality," paper presented at the American Public Works Assocation Urban Drainage Workshop, Omaha, Nebraska (March 14, 1978).

MACHMEIER, R. E. and L. Larson. "Runoff Hydrographs for Mathematical Watershed Model," *J. Hyd. Div.*, ASCE, 94(HY6):1453–1474 (1968).

MEIN, R. G. and C. L. Larson. "Modeling Infiltration During a Steady Rain," *Water Resources Res.*, 9(2):384–394 (1973).

MITCHELL, J. and B. A. Jones, Jr. "Micro-Relief Surface Depression Storage: Analysis of Models to Describe the Depth-Storage Function," *Water Resources Bull.*, 12(6):1205–1222 (1976).

MOREL-SEYTOUX, H. J. and J. Khanji. "Derivation of an Equation of Infiltration," *Water Resources Res.*, 10(4):795–800 (1974).

MORGAN, P. E. and S. M. Johnson. "Analysis of Synthetic Unit-Graph Methods," *J. Hyd. Div.*, ASCE, 88(HY5):199–220 (1962).

NASH, J. E. "Systematic Determination of Unit Hydrograph Parameters," *J. Geophysical Res.*, 64 (1):111–115 (1959).

O'LOUGHLIN, E. M. "Modelling Water Quality," in *Prediction in Catchment Hydrology*. T. G. Chapman and F. X. Dunin, eds. Netly, S. Australia:The Griffin Press, pp. 337-355 (1975).

OVERTON, D. E. and M. E. Meadows. *Stormwater Modeling*. New York:Academic Press, Inc., p. 159 (1976).

PAPADAKIS, C. N. and H. C. Preul. "Testing of Methods for Determination of Urban Runoff," *J. Hyd. Div.*, ASCE (HY9):1319-1335 (1973).

PILGRIM, D. H. "Model Evaluation, Testing and Parameter Estimation in Hydrology," in *Prediction in Catchment Hydrology*. T. G. Chapman and F. X. Dunin, eds. Netly, S. Australia:The Griffin Press, pp. 305-336 (1975).

PITT, R. and R. Field. "Water Quality Effects From Urban Runoff," *J. Am. Water Works Assoc.*, 69(8):432-436 (1977).

PRASAD, R. "Nonlinear Simulation of a Regional Hydrologic System," Ph.D. Thesis, University of Illinois, Urbana, Illinois (1965).

RAGAN, R. M. and M. J. Root. "A Linked System Model for the Synthesis of Hydrographs in Urban Areas," A Report to the Maryland State Highway Administration, Annapolis, Maryland, and the Federal Highway Administration, Washington, DC (1974).

REED, D. W. et al. "A Non-Linear Rainfall-Runoff Model Providing for Variable Lag Time," *J. Hydrology*, 25(3/4):295-305 (1975).

ROBERTS, F. S. *Discrete Mathematical Models*. Englewood Cliffs, NJ:Prentice-Hall, Inc., 559 pp. (1976).

ROBERTSON, A. F. et al. "Runoff from Impervious Surfaces Under Conditions of Simulated Rainfall," *Trans. Am. Soc. Agric. Eng.*, 9(3):343-346 (1966).

SARTOR, J. D. and G. B. Boyd. "Water Pollution Aspects of Street Surface Contaminants," U.S. Environmental Protection Agency, Report No. EPA-R2-72-081, Washington, DC, 236 pp. (1972).

SARTOR, J. D. and G. B. Boyd. "Water Pollutants in Urban Runoff: A Summary of Water Pollution Aspects of Street Surface Contaminants," U.S. Environmental Protection Agency Report No. EPA-R2-72-81, Washington, DC (1977).

SCHAAKE, J. C., JR. et al. "Experimental Examination of the Rational Method," *J. Hyd. Div.*, ASCE (HY6):353-370 (1967).

SHERMAN, L. K. "Streamflow from Rainfall by the Unit-Graph Method," *Eng. News Rec.*, 108: 401-505 (1932).

SHERMAN, L. K. "The Unit Hydrograph Method," in *Hydrology*. O. Meinzer, ed. New York: McGraw-Hill Book Co., Inc., pp. 514-525 (1942).

SINGH, K. P. "Nonlinear Instantaneous Unit-Hydrograph Theory," *J. Hyd. Div.*, ASCE, 90 (HY2): 313-323 (1964).

SINGH, V. P. and D. A. Woolhiser. "A Nonlinear Kinematic Wave Model for Watershed Surface Runoff," *J. Hydrology*, 31:221-243 (1976).

SINGH, V. P. and S. Buapeng. "Effect of Rainfall-Excess Determination on Runoff Computation," *Water Resources Bull.*, 13(3):499-514 (1977).

SINGH, V. P. and Y. K. Birsoy. "Some Statistical Relationships Between Rainfall and Runoff," *J. Hydrology*, 34:251-268 (1977).

SNYDER, W. and J. Stall. "Men, Models, Methods and Machines in Hydrologic Analysis," *J. Hyd. Div.*, ASCE, 91(HY2):85-99 (1965).

SPIEGEL, M. R. "Probability and Statistics" *Schaum's Outline Series in Mathematics*. New York: McGraw-Hill Book Company, 372 pp. (1975).

SUTHERLAND, R. C. and R. H. McCuen. "Simulation of Urban and Nonpoint Source Pollution," *Water Resources Bull.*, 14(2):409-428 (1978).

TAYLOR, G. I. "The Dispersion of Matter in Turbulent Flow Through a Pipe," *Proc. Roy. Soc. London*, 234A:67 (1954).

TAYLOR, A. B. and H. E. Schwarz. "Unit-Hydrograph Lag and Peak Flow Related to Basin Characteristics," *Trans. Am. Geophys. Union*, 32:235-246 (1952).

VIESSMAN, W., JR. et al. "Urban Storm Runoff Relations," *Water Resources Res.*, 6(1):275-279 (1970).

VIESSMAN, W., JR. et al. "Runoff Estimation for Very Small Drainage Areas," *Water Resources Res.*, 4(1):87-93 (1968).

VIESSMAN, W., JR. et al. *Introduction to Hydrology*. New York:Harper and Row, Publishers, Inc., 704 pp. (1977).

VIESSMAN, W., JR. et al. "The Hydrology of Small Impervious Areas," *Water Resources Res.*, 2(3):405-412 (1966).

WATT, W. E. and C. H. R. Kidd. "QUURM—A Realistic Urban Runoff Model," *J. Hydrology*, 27:225-235 (1975).

WILLEKE, G. E. "Time in Urban Hydrology," *J. Hyd. Div.*, ASCE, 92(HY1):13-29 (1966).

WILLIS, R. et al. "Steady-State Water Quality Modeling in Streams," *J. Env. Eng. Div.*, ASCE, (EE2):245-258 (1975).

WOLMAN, M. G. and J. P. Miller. "Magnitude and Frequency of Forces in Geomorphic Processes," *J. Geol.*, 68:54-74 (1960).

WOOLHISER, D. A. and J. A Liggett. "Unsteady, One-Dimensional Flow Over a Sloping Plane—The Rising Hydrograph," *Water Resources Res.*, 3(3):753-771 (1967).

YALIN, M. S. "An Expression for Bedload Transportation," *J. Hyd. Div.*, ASCE, 89(HY3):221-250 (1963).

YOTSUKURA, N. and W. W. Sayre. "Transverse Mixing in Natural Channels," *Waste Resources Res.*, 12(4):694-704 (1976).

Selected Readings

Aitken, A. P. "Catchment Models for Urban Areas," in *Prediction in Catchment Hydrology*. T. G. Chapman and F. X. Dunin, eds. Netly, S. Australia:The Griffin Press, pp. 257-275 (1975).

Akan, A. O. "Inlet Concentration Time Nomogram for Urban Basins," *Water Resources Bulletin*, 20(2):267-670 (1984).

Alley, W. M. et al. "Parametric-Deterministic Urban Watershed Model," *Journal of Hyraulics Division*, American Society of Civil Engineers, 106(HY5):679-690 (1980).

Ames, W. F. *Nonlinear Problems of Engineering*. New York:Academic Press, Inc. (1964).

Amorocho, J. "Deterministic Nonlinear Hydrologic Models," in *The Progress of Hydrology. Vol. 1: New Developments in Hydrology*. Proceedings of the First International Seminar for Hydrology Professors, University of Illinois, Urbana, Illinois, July 13-26, 1969. pp. 420-472 (1969).

Amorocho, J. and W. E. Hart. "A Critique of Current Methods of Hydrologic Systems Investigation," *Trans. Am. Geophys. Union*, 45:307-321 (1964).

Ardis, C. V. et al. "Storm Drainage Practices of Thirty-Two Cities," *J. Hyd. Div.*, ASCE, No. HY1, pp. 383-408 (1969).

Arnell, V. "Estimating Runoff Volumes from Urban Areas," *Water Resources Bulletin*, 18(3):383-387 (1982).

Artken, A. P. "Assessing Systematic Errors in Rainfall–Runoff Models," *J. Hydrology*, 20(2): 121–136 (1973).

Batty, M. *Urban Modelling: Algorithms, Calibrations, Predictions.* New York:Cambridge University Press, 406 pp. (1976).

Bayer, M. B. "A Modeling Method for Evaluating Water Quality Policies in Nonserial River Systems," *Water Resources Bull.*, 13(6):1141–1152 (1977).

Beard, L. R. "Optimization Techniques for Hydrologic Engineering," *Water Resources Res.*, 3(3):809–815 (1967).

Beaudoin, P., J. Rousselle and G. Marchi. "Reliability of the Design Storm Concept in Evaluating Runoff Peak Flow," *Water Resources Bulletin*, 19(3):483–484 (1983).

Benjes, H. H. et al. "Stormwater Overflows from Combined Sewers," *J. Water Poll. Control Fed.*, 33(12) (1961).

Benzie, W. J. and R. J. Courchaine. "Discharges from Separate Storm Sewers and Combined Sewers," *J. Water Poll. Control Fed.*, 38(3):410–421 (1966).

Bernard, M. "A Modified Rational Method of Estimating Flood Flows," in *Low Dams—A Manual for Small Water Storage Projects*. Washington, DC:National Res. Comm. (1938).

Betson, R. P. and R. F. Green. "Analytically Derived Unit Graph and Runoff," *J. Hyd. Div.*, ASCE, 94(HY6):1489–1505 (1968).

Betson, R. P. and J. B. Marius. "Source Areas of Storm Runoff," *Water Resources Res.*, 5(3): 574–582 (1969).

Body, D. N. "Empirical Methods and Approximations in the Determination of Catchment Response," in *Prediction in Catchment Hydrology*. T. G. Chapman and F. X. Dunin, eds. Netly, S. Australia:The Griffin Press, pp. 293–304 (1975).

Brater, E. F. "Steps Toward a Better Understanding of Urban Runoff Processes," *Water Resources Res.*, 4(2):335–347 (1968).

Burm, R. J. et al. "Chemical and Physical Comparison of Combined and Separate Sewer Discharges," *J. Water Poll. Control Fed.*, 40(1):112–126 (1968).

Burm, R. J. and R. D. Vaughan. "Bacteriological Comparison Between Combined and Separate Sewer Discharges in Southeastern Michigan," *J. Water Poll. Control Fed.*, 38(3):400–409 (1966).

Butler, S. S. *Engineering Hydrology.* Englewood Cliffs, NJ:Prentice-Hall, Inc., 356 pp. (1957).

Chery, D. L. "Design and Test of a Physical Watershed Model," *J. Hydrology*, 4:224–235 (1966).

Chien, J. S. and K. K. Saigal. "Urban Runoff by Linearized Subhydrograph Method," *J. Hyd. Div.*, ASCE, 100(HY8):1141–1157 (1974).

Chow, V. T. and S. J. Kareliotis. "Analysis of Stochastic Hydrologic Systems," *Water Resources Res.*, 6(6):1569–1582 (1970).

Ciriani, T. A. et al. *Mathematical Models for Surface Water Hydrology.* New York:John Wiley and Sons, Inc., 423 pp. (1977).

Clark, C. O. "Storage and the Unit Hydrograph," *ASCE Trans.*, Paper No. 2261, 72(110): 1419–1447 (1943).

Colston, N. V., Jr. "Characterization and Treatment of Urban Land Runoff," National Environmental Research Center, Office of Research and Development, U.S. Environmental Protection Agency, Cincinnati, Ohio, EPA Technology Series No. EPA-670/2-74-096 (1974).

Crawford, N. H. "Practical Applications of Hydrologic Simulation," in *Systems Analysis of Hydrologic Problems*, NSF Seminar, August 2–14, 1970, pp. 326–342 (1970).

Crawford, N. H. and R. F. Linsley. "The Synthesis of Continuous Streamflow Hydrographs on a Digital Computer," Technical Report No. 12, Stanford, California, Stanford University, Department of Civil Engineering (1962).

Cruise, J. F. and D. N. Contractor. "Unit Hydrographs for Urbanizing Watersheds," *Journal of Hydraulics Division*, American Society of Civil Engineers, 106(HY3)440–445 (1980).

D'Angelo, H. *Linear Time-Varying Systems: Analysis and Synthesis.* Boston:Allyn and Bacon, 346 pp. (1970).

Dawdy, D. R. and T. O'Donnell. "Mathematical Models of Catchment Behavior," *J. Hyd. Div.*, ASCE, 91(HY4):123–137 (1965).

Dawdy, D. R. and R. W. Lichty. "Methodology of Hydrologic Model Building," International Association of Scientific Hydrology, Publication No. 81, AIHS, pp. 347–355 (1968).

Dawdy, D. R. "Mathematical Modeling in Hydrology," in *The Progress of Hydrology. Vol. 1:New Developments in Hydrology.* Proceedings of the First International Seminar for Hydrology Professors, University of Illinois, Urbana, Illinois, July 13–26, pp. 346–361 (1969).

Dawdy, D. R. et al. "A Rainfall–Runoff Simulation Model for Estimation of Flood Peaks for Small Drainage Basins," U.S. Geological Survey Professional Paper No. 506-B (1972).

Dawdy, D. R. and G. P. Kalivin. "Mathematical Modeling in Hydrology," *Bull. Int. Assoc. Sci. Hyd.*, 15(1) (1970).

Debo, T. N. "Urban Flood Damage Estimating Curves," *Journal of Hydraulics Division*, American Society of Civil Engineers, 108(HY10):1059–1069 (1982).

Defilippi, J. A. and C. S. Shih. "Characteristics of Separated Storm and Combined Sewer Flows," *J. Water Poll. Control Fed.*, 43(10):2033–2058 (1971).

Diskin, M. H. "A Basic Study of the Linearity of the Rainfall–Runoff Process in Watersheds," Ph.D. Thesis, University of Illinois, Urbana, Illinois (1964).

Diskin, M. H. "Definition and Uses of the Linear Regression Model," *Water Resources Res.*, 6 (6):1668–1673 (1970).

Di Toro, D. M. "Probability Model of Stream Quality Due to Runoff," *Journal of Environmental Engineering Division*, American Society of Civil Engineers, 110(3):607–628 (1984).

Dobbins, W. E. "BOD and Oxygen Relationships in Streams," *J. San. Eng. Div.*, ASCE, 90 (SA3):53–78 (1964).

Dooge, J. C. I. "Problems and Methods of Rainfall–Runoff Modeling," in *Mathematical Models for Surface Water Hydrology*. T. A. Ciriani et al., eds. New York:John Wiley and Sons, Inc., pp. 71–108 (1977).

Dooge, J. C. I. "Linear Theory of Hydrologic Systems," Technical Bulletin No. 1468, Agricultural Research Service, U.S. Department of Agriculture, Washington, DC, 327 pp. (1973).

Dracup, J. A. et al. *Synthesis and Evaluation of Urban-Regional Hydrologic Rainfall–Runoff Criteria.* Los Angeles:Environmental Dynamics, Inc. (1973).

Dunbar, D. D. and J. G. F. Henry. "Pollution Control Measures for Stormwaters and Combined Sewer Overflows," *J. Water Poll. Control. Fed.*, 38(1):9–25 (1966).

Eagleson, P. S. "Deterministic Linear Hydrologic Systems," in *The Progress of Hydrology, Vol. I: New Developments in Hydrology.* Proceedings of the First International Seminar for Hydrology Professors, University of Illinois, Urbana, Illinois, July 13–26, 1969. pp. 400–419 (1969).

Eagleson, P. S. "Potential of Physical Models for Achieving Better Understanding and Evaluation of Watershed Changes,' 'in *Effects of Watershed Changes on Streamflow*. W. L. Moore and C. W. Morgan, eds. Austin:University of Texas Press, pp. 12–25 (1969).

Ellis, F. W. et al. "Converging Flow Model Applied to Urban Catchment," *Journal of Hydraulics Division*, American Society of Civil Engineers, 106(HY9)1457–1470 (1980).

Selected Readings

Espey, W. H., Jr. et al. "Urban Effects on the Unit Hydrograph, in *Effects of Watershed Changes on Streamflow*. W. L. Moore and C. W. Morgan, eds. Austin:University of Texas Press, pp. 215–228 (1969).

Federal Water Quality Administration, Department of the Interior. "Storm Water Pollution from Urban Land Activity," Water Pollution Control Research Series, Program No. 11034, Contract No. 14-12-187 (1970).

Fetterolf, C. M., Jr. "Mixing Zone Concepts," in *Biological Methods for the Assessment of Water Quality*. Philadelphia:The American Society for Testing and Materials, pp. 31–45 (1969).

Fogel, M. M. et al. "Choosing Hydrologic Models for Management of Changing Watershed," in *Watersheds in Transition*. S. C. Csallany et al., eds. Urbana, IL:AWRA Publications, pp. 118–123 (1972).

Frind, E. O. "Rainfall-Runoff Relationships Expressed by Distribution Parameters," *J. Hydrology*, 9:405–426 (1969).

Fruh, E. G. "Urban Effects on Quality of Streamflow," in *Effects of Watershed Changes on Streamflow*. W. L. Moore and C. W. Morgan, eds. Austin:University of Texas Press, pp. 255–282 (1969).

Glenne, B. "Simulation of Water Pollution Generation and Abatement on Suburban Watersheds," *Water Resources Bulletin*, 20(2):211–217 (1984).

Glymph, L. M. et al. "Hydrologic Response of Watersheds to Land Use Management," *J. Irr. & Drainage Div.*, ASCE (IR2):305–318 (1971).

Graham, P. H. et al. "Estimation of Imperviousness and Specific Curb Length for Forecasting Stormwater Quality and Quantity," *J. Water Poll. Control Fed.*, 46(4):717–725 (1974).

Gray, D. M. "Synthetic Unit Hydrographs for Small Watersheds," *J. Hyd. Div.*, ASCE, 87(HY4): 33–54 (1961).

Gregory, R. L. and C. E. Arnold. "Runoff—Rational Runoff Formulas," *Trans. ASCE*, 96: 1038–1099 (1932).

Grigg, N. S. and T. Sriburi. "Order Classification of Urban Catchments," *Water Resources Bull.*, 14(1):63–71 (1978).

Gupta, V. L. et al. "Analytical Modeling of Surface Runoff Hydrographs for Major Streams in Northeast Thailand," *Hyd. Sci. Bull.*, 19(4):523–540 (1974).

Heeps, D. P. and R. G. Mein. "Independent Comparison of Three Urban Runoff Models," *J. Hyd. Div.*, ASCE, 100(HY7):995–1009 (1974).

Heerdegen, R. G. and B. M. Reich. "Unit Hydrographs for Catchments of Different Sizes and Dissimilar Regions," *J. Hydrology*, 22:143–153 (1974).

Himmelblau, D. M. *Applied Nonlinear Programming*. New York:McGraw-Hill Book Company, 498 pp. (1972).

Holtan, H. N. "USDAHL-74 Model of Watershed Hydrology," U.S. Department of Agriculture Technical Bulletin, 110 pp. (1974).

Hornberger, G. M. et al. "The Relationship Between Light and Photosynthetic Rate in a River Community and Implications for Water Quality Modeling," *Water Resources Res.*, 12(4): 723–730 (1976).

Horton, R. E. "Surface Runoff Phenomena: Part I, Analysis of the Hydrograph," Horton Hydrological Laboratory Publication No. 101, Ann Arbor MI:Edwards Bros. Inc. (1935).

Huber, W. C. et al. "Storm Water Management Model, User's Manual," prepared for the Environmental Protection Agency by the Department of Environmental Engineering Sciences, University of Florida, Gainesville, Florida, EPA-670/2-75-017, 367 pp. (1975).

Ibbitt, R. P. and T. O'Donnell. "Fitting Methods for Conceptual Catchment Models," *J. Hyd. Div.*, ASCE, 97(HY9):1331–1342 (1971).

Johnson, C. F. "Equipment, Methods and Results from Washington, D.C. Sewer Overflow Studies," *J. Water Poll. Control Fed.*, 33(7):721 (1961).

Kaltenbach, A. B. "Storm Sewer Design by the Inlet Method," *Public Works*, 94(1):86–89 (1963).

Kidd, C. H. R. and P. R. Helliwell. "Simulation of the Inlet Hydrograph for Urban Catchments," *J. Hydrology*, 35(1/2):159–172 (1977).

Kisiel, C. C. "Mathematical Methodology in Hydrology," in *The Progress of Hydrology: Vol. 1, New Developments in Hydrology*, Proceedings of the First International Seminar for Hydrology Professors, University of Illinois, Urbana, Illinois, July 13–26, 1969, pp. 362–399 (1969).

Koivo, A. J. and G. Phillips. "Optimal Estimation of DO, BOD and Stream Parameters Using a Dynamic Discrete Time Model," *Water Resources Res.*, 12(4):705–711 (1976).

Kothandaraman, V. and B. B. Ewing. "A Probabilistic Analysis of Dissolved Oxygen–Biochemical Oxygen Demand Relationship in Streams," *J. Water Poll. Control Fed.*, 41(2):R73–R90 (1969).

Labadie, T. W. and J. A. Dracup. "Optimal Identification of Nonlinear Hydrologic System Response Models by Quasilinearization," *Water Resources Res.*, 5(3):583–590 (1969).

Lane, L. J. and D. A. Woolhiser. "Simplifications of Watershed Geometry Affecting Simulation of Surface Runoff," *J. Hydrology*, 35(1/2):173–190 (1977).

Laurenson, E. M. **"A Catchment Storage Model for Runoff Routing,"** *J. Hydrology*, 2:141–163 (1964).

Laurenson, E. M. "Streamflow in Catchment Modeling," in *Prediction in Catchment Hydrology*. C. Chapman and F. Dunin, eds. Netley, S. Australia:Australian Academy of Science, pp. 149–164 (1975).

Lefeuvre, A. R. "Use of Computer Technology to Develop Mathematical Models for Natural Bodies," in *Water and Water Pollution Handbook, Vol. 1*. L. L. Ciaccio, ed. New York:Marcel Dekker, Inc., pp. 263–295 (1971).

Linsley, R. K. "A Critical Review of Currently Available Hydrologic Models for Analysis of the Urban Stormwater Runoff," Department of the Interior, Office of Water Resources Research, Washington, DC (1971).

Lombardo, P. S. and D. D. Franz. "Mathematical Model of Water Quality in Rivers and Impoundments," Palo Alto, CA:Hydrocomp, Inc. (1972).

Maksimovic, C. and M. Radojkovic, eds. *Urban Drainage Modelling*. Oxford, England:Pergammon Press, Ltd., 450 pp. (1986).

Mandeville, A. N. and T. O'Donnell. "Introduction of Time Variance to Linear Conceptual Catchment Models," *Water Resources Res.*, 9(2):298–310 (1973).

Marsalek, J. et al. "Comparative Evaluation of Three Urban Runoff Models," *Water Resources Bull.*, 11(2):306–328 (1975).

Matalas, N. C. "Statistics of a Runoff–Precipitation Relation," U.S. Geological Survey Professional Paper 434-D, 9 pp. (1963).

McCuen, R. H. et al. "Estimating Urban Time of Concentration," *Journal of Hydraulics Division*, American Society of Civil Engineers, 110(7):887–904 (1984).

McCuen, R. H. et al. "SCS Urban Peak Flow Methods," *Journal of Hydraulics Division*, American Society of Civil Engineers, 110(HY3):290–299 (1984).

Metcalf and Eddy, Inc. "Storm Water Management Model, Environmental Protection Agency, Vol. I," University of Florida, Water Resources Engineers, Inc. (1971).

Minshall, N. E. "Predicting Storm Runoff on Small Experimental Watersheds," *J. Hyd. Div.*, ASCE, 86(HY8):17–38 (1960).

Mitchell, W. D. "Unit Hydrographs in Illinois," U.S. Department of the Interior, U.S. Geological Survey, Division of Waterways, State of Illinois, 294 pp. (1948).

Moore, I. D. and R. G. Mein. "Evaluating Rainfall–Runoff Model Performance," *J. Hyd. Div.*, ASCE, 102(HY9):1390–1395 (1976).

Morel-Seytoux, H. J. "Derivation of Equations for Rainfall Infiltration," *J. Hydrology*, 31: 203–219 (1976).

Morgali, J. R. and R. K. Linsley. "Computer Analysis of Overland Flow," *J. Hyd. Div.*, ASCE, 91(HY3):81–100 (1964).

Murphy, J. B. et al. "Geomorphic Parameters Predict Hydrograph Characteristics in the Southwest," *Water Resources Bull.*, 13(1):25–38 (1977).

Muzik, I. "Laboratory Experiments with Surface Runoff," *J. Hyd. Div.*, ASCE, 100(HY4): 501–516 (1974).

Narayana, D. V. et al. "Analog Computer Simulation of the Runoff Characteristics of an Urban Watershed," Research Project Technical Completion Report to Office of Water Resources Research, Department of the Interior, December 1968, Washington, DC, 83 pp. (1968).

Narayana, D. V. et al. "Simulation of Runoff from Urban Watersheds," *Water Resources Bull.*, 7 (1):54–68 (1971).

Nash, J. E. "The Form of the Instantaneous Unit Hydrograph," International Association of Scientific Hydrology Publication No. 42, pp. 114–118 (1957).

Natale, L. and E. Todini. "A Constrained Parameter Estimation Technique for Linear Models in Hydrology," in *Mathematical Models for Surface Water Hydrology*. T. A. Ciriani, et al., eds. New York:John Wiley & Sons, Inc., pp. 109–147 (1977).

Natale, L. and E. Todini. "A Stable Estimation for Linear Models: 2. Real-World Hydrologic Applications," *Water Resources Res.*, 12(4):672–676 (1976).

Onstad, C. A. and D. G. Jamieson. "Modeling the Effect of Land Use Modifications on Runoff," *Water Resources Res.*, 6(5):1287–1295 (1970).

Overton, D. E. and D. L. Brakensiek. "A Kinematic Model of Surface Runoff Response," *Proceedings*, Symposium Results Res. Representative Exptl. Basins, Wellington, New Zealand, December, 1970. 1. IASH-UNESCO, Paris (1973).

Phamwon, S. and Y.-S. Fok. "Urban Runoff Digital Computer Model," *J. Hyd. Div.*, ASCE, No. 13068 (HY7):723–735 (1977).

Porter, J. W. and T. A. McMahon. "A Model for the Simulation of Streamflow Data from Climatic Records," *J. Hydrology*, 13:297–324 (1971).

Prasad, R. "A Nonlinear Hydrologic System Response Model," *J. Hyd. Div.*, ASCE, 93(HY4): 201–221 (1967).

Preul, H. C. and C. N. Papadakis. "Urban Runoff Characteristics," Report No. 11024, DQU, Environmental Protection Agency Water Quality Office, Water Pollution Control Research Series, Washington, DC (1972).

Ragan, R. M. "Parameter Estimates for Urban Hydrologic Models Using Computer Aided Analysis of Satellite Data," paper presented at the Spring Annual Meeting, American Geophysical Union, Washington, DC (1975).

Rao, R. A. and J. W. Delleur. "Instantaneous Unit Hydrographs, Peak Discharges and Time Lags in Urban Basins," *Hydrological Sci. Bull.*, 19(2):185–198 (1974).

Rao, R. A. et al. "Conceptual Hydrologic Models for Urbanizing Basins," *J. Hyd. Div.*, ASCE (HY8):1205–1220 (1972).

Rao, R. A. "Nonlinear Analysis of the Rainfall–Runoff Process," Ph.D. Thesis, University of Illinois, Urbana, Illinois (1968).

Riley, J. P. "Computer Simulation of Water Resource Systems," in *Systems Analysis of Hydrologic Problems*, National Science Foundation Seminar, August 2-14, 1970, pp. 249-274 (1970).

Roesner, L. A. et al. "Use of Storm Drainage Models in Urban Planning," in *Watersheds in Transition*. S. C. Csallany et al., eds. Urbana, IL:AWRA Publications, pp. 400-405 (1972).

Rosendahl, P. C. and T. D. Waite. "Transport Characteristics of Phosphorus in Channelized and Meandering Streams," *Water Resources Bull.*, 14(5):1227-1238 (1978).

Rovey, E. W. and D. A. Woolhiser. "Urban Storm Runoff Model," *J. Hyd. Div.*, ASCE, (HY11):1339-1351 (1977).

Russel, S. O. et al. "Estimating Design Flows for Urban Drainage," *Journal of Hydraulics Division*, American Society of Civil Engineers, 105(HY1):43-52 (1979).

Sarma, P. B. S. et al. "Comparison of Rainfall-Runoff Models for Urban Areas," *J. Hydrology*, 18(3/4):329-347 (1973).

Schneider, W. J. "Precipitation as a Variable in the Correlation of Runoff Data," U.S. Geological Survey Professional Paper 424-B, pp. 20-21 (1961).

Schneider, W. J. "Water Data for Metropolitan Areas," U.S. Geological Survey Water Supply Paper 1871, 397 pp. (1968).

Seaburn, G. E. "Method of Rating Flow in a Storm Sewer," U.S. Geological Survey Professional Paper 750-D, pp. D219-D223 (1971).

Shubinski, R. P. and L. A. Roesner. "A Mathematical Model of Urban Storm Drainage," in *Systems Analysis of Hydrologic Problems*, National Science Foundation Seminar, August 2-14, 1970, pp. 379-400 (1970).

Singh, K. "Unit Hydrographs, A Comparative Study," *Water Resources Bull.*, 12(2):381-392 (1976).

Singh, V. P. "A Rapid Method of Estimating Mean Areal Rainfall," *Water Resources Bull.*, 12(2):307-316 (1976).

Singh, V. P. "Sensitivity of Some Runoff Models to Errors in Rainfall Excess," *J. Hydrology*, 33:301-318 (1977).

Singh, V. P. "Hybrid Formulation of Kinematic Wave Models of Watershed Runoff," *J. Hydrology*, 27(1/2):33-50 (1975).

Singh, V. P. "Kinematic Wave Modeling of Watershed Surface Runoff: A Hybrid Approach," International Association for Hydrologic Sciences Publication No. 117, pp. 255-264 (1975).

Smith, J. M. *Mathematical Modeling and Digital Simulation for Engineers and Scientists*. New York:John Wiley and Sons, Inc., 332 pp. (1977).

Smith, R. E. and D. A. Woolhiser. "Overland Flow on an Infiltrating Surface," *Water Resources Res.*, 7(4):899-913 (1971).

Stall, J. B. and M. L. Terstriep. "Storm Sewer Design: An Evaluation of the RRL Method," Environmental Protection Agency, Office of Research and Monitoring, Washington, DC, EPA-R2/72-068, 73 pp. (1972).

Tang, W. H. et al. "Optimal Risk-Based Design of Storm Sewer Network," *J. Env. Eng. Div.*, ASCE, 101(EE3):381-398 (1975).

Terstriep, M. L. and J. B. Stall. "The Illinois Urban Drainage Area Simulator, ILLUDAS," Bulletin 58, Illinois State Water Survey, Urbana, Illinois (1974).

Terstriep, M. L. "Urban Runoff by Road Research Laboratory Methods," *J. Hyd. Div.*, ASCE, 95 (HY6):1809-1834 (1969).

Tholin, A. L. and C. J. Keifer. "Hydrology of Urban Runoff," *J. San. Eng. Div.*, ASCE, 125:1308-1351 (1959).

U.S. Public Health Services. "Pollutant Effects of Stormwater Overflows," USPHS Publication No. 1246 (1964).

Selected Readings

Van Sickle, D. "Experience with the Evaluation of Urban Effects for Drainage Design," in *Effects of Watershed Changes on Streamflow*. W. L. Moore and C. W. Morgan, eds. Austin: University of Texas Press, pp. 229-254 (1969).

Verhoff, F. H. et al. "Modeling of Nutrient Cycling in Microbial Aquatic Ecosystems; Theoretical Considerations," in *The Aquatic Environment*, Symposium sponsored by the Environmental Protection Agency, Office of Water Program Operations, Washington, DC, pp, 13-56 (1972).

Viessman, W., Jr. "Assessing the Quality of Urban Drainage," *Public Works*, 100(10):88-92 (1969).

Wallis, J. R. and E. Todini. "Comment on the Residual Mass Curve Coefficient," *J. Hydrology*, 24(3/4)201-205 (1975).

Weber, J. E. et al. "On the Mismatch Between Data and Models of Hydrologic and Water Resource Systems," *Water Resources Bull.*, 7(6):1075-1089 (1973).

Whisler, F. D. and H. Bouwer. "Comparison of Methods for Calculating Vertical Drainage and Infiltration for Soils," *J. Hydrology*, 10(1):1-19 (1970).

Willeke, G. E. "The Prediction of Runoff Hydrographs for Urban Watersheds from Precipitation Data and Watershed Characteristics," Abstract, *J. Geophys. Res.*, 67(9):3610 (1962).

Willis, R. et al. "Transient Water Quality Modeling in Streams," *Water Resources Bull.*, 12(1): 157-174 (1976).

Wooding, R. A. "A Hydraulic Model for the Catchment-Stream Problem," *J. Hydrology*, 3(3/4): 254-267 (1965).

Wooding, R. A. "A Hydraulic Model for the Catchment-Stream Problem: 3) Comparison with Runoff Observations," *J. Hydrology*, 4(1):21-37 (1966).

Wu, J. S. and R. C. Ahlert. "Assessment of Methods for Computing Storm Runoff Loads," *Water Resources Bull.*, 14(2):429-439 (1978).

Yen, B. C. and A. S. Sevuk. "Design of Storm Sewer Networks," *J. Env. Eng. Div.*, ASCE (EE4):535-553 (1975).

Yoon, Y. N. and H. G. Wenzel, Jr. "Mechanics of Sheet Flow under Simulated Rainfall," *J. Hyd. Div.*, ASCE, 97(HY9):1367-1386 (1971).

6 Nonstructural Control Measures

> ... as our urban areas continue to expand and coalesce into metropolitan areas, the cost, damage, and environmental and disruptive effects of floods become intolerable; and yet the very growth of cities complicates the problem of finding adequate space for storm drainage systems. The lack of adequate planning, especially with regard to environmental impact, may well produce disastrous results.
> —Espey and Winslow (1974), p. 279.

6.1 Planning and Planning Commissions

PLANNING

SIMON ET AL. (1950) defined planning[50] as:

> ... that activity that concerns itself with proposals for the future, with the evaluation of alternative proposals, and with the methods by which these processes may be achieved. Planning is rational, adaptive thought applied to the future, and to matters over which the planners or the administrative organizations with which they are associated have some degree of control (pp. 423-424).

As Ewing (1969) pointed out:

> ... the term "planning" is an old one. It has been in many languages for a great many centuries—ever since man began thinking of the future implications of current choices of action (p. 3).

He goes on to state further:

> ... in times past, planning could be considered a luxury, a useful "extra" in the arsenal of tools used by leaders seeking to bring about change. Today, this is no longer so. We seem to be fast approaching a point at which the very fate of corporations, cities, public service agencies, military organizations, state governments, and even whole nations and populations will depend on their leaders' willingness and ability to plan (p. 3).

Ewing suggests that the industrial revolution, the population explosion and certain western values concerning progress and change have made planning a

---

[50]Seeley (1962) and Walker (1950) spent several pages each in developing explicit definitions for the word "planning." These pages make interesting reading and provide a good overall definition.

necessity. The crucial reason for this is that organizations depend greatly on one another. He writes further:

> ... once institutions like big governments, big labor, big agriculture, big business and big health and welfare become interdependent, once they begin dividing into numerous subinstitutions—the many agencies in Washington; the chains of national, regional and local offices in labor; the networks of growers, processors—then planning becomes a condition of existence. The big question in planning becomes not whether it is justified, but to what extent and in what manner it shall be practiced (p. 4).

Claire (1973) noted that in 1960 there were over 91,000 governmental units in the United States, two-thirds of which were single-function units, such as sanitary districts, school boards, etc. Since that time, he estimates that the number has grown to well over 100,000. There are more than 35,000 township and municipal governments, and 3,049 county governments, as well as other types of local governments such as water districts and housing authorities. Most of the 35,000 units practice formal planning in one form or another, although they might not have formal planning commissions.

THE PLANNING COMMISSION

Lewis (1949) reported that the states of New York and New Jersey were the first in the United States to adopt legislation authorizing cities to appoint planning commissions; this occurred in 1913. In 1926, New York state initiated enabling acts for municipalities to engage in comprehensive planning. Since then, the remaining states have adopted similar enabling acts, and a multitude of planning commissions have been established.[51]

Benckert (1973) reported that the cost of fundamental planning represents from 0.5 to 1.5% of capital expenditures. He pointed out that these figures would hold true for organizations that have, in the past, had a commitment to planning and have developed the expertise. Communities lacking a long-range plan and such a prior commitment must face up to a major effort and expenditure to establish such planning capabilities.

Funding for planning commissions may come from many sources. The local governmental unit(s) usually supplies the bulk of the money, and various federal agencies such as the Department of Housing and Urban Development (1954 Housing Act), Department of Transportation, Department of Health, Education and Welfare, the Environmental Protection Agency, the Law Enforcement Assistance Administration, and possibly others may match municipal funds or supply funding in return for planning services. In all cases, the costs of planning are minimal when compared to the chronic expenditures which usually accompany the lack of planning.

---

[51] For an excellent history of early planning movements, see Walker (1950).

PLANNING COMMISSION STRUCTURE

Benckert (1973) wrote:

> . . . organization of a Planning Agency varies with the size of a city. A smaller one (under 10,000 population) usually has no staff, a medium-size city (10,000 to 100,000) may have a Planning Director and a small staff, and a large city (100,000 or more) may have an Executive Director with a Director of Planning, Director of Public Relations, and a Director of Budget (p. 367).

The structure of a planning commission (planning agency, planning board, citizens committee or citizens action commission) is usually dictated by the legislation enabling the establishment of the planning group (see Figure 6.1). Claire (1973) reports that planning commissions are usually appointed by the mayor and confirmed by the city council, and customarily serve without pay as a civic duty. Many small communities have such a lay group consisting of concerned citizens. These planning commissions do not employ professional planners.

Planning commissions serving larger communities still retain and maintain the basic core commission of local citizen representatives, but usually find

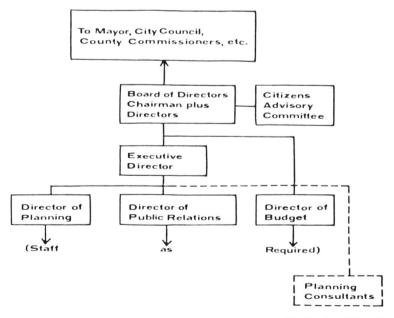

Figure 6.1. Organization of planning agency by structure (*Source*: *Handbook on Urban Planning*. By W. H. Claire, © 1973 by Litton Educational Publishing, Inc., reprinted by permission of Van Nostrand Reinhold Company).

that they need a professional planning staff. The planning staff handles the routine matters (i.e., advises the client community), advises the commission and is in turn advised by the commission.

Larger communities and regions usually find that their planning commission is too large to meet on a regular basis, and elect or select an executive committee or board of directors to meet frequently. These individuals are usually much more professional in their approach to planning in comparison to the lay member planning commission members described above. Benckert (1973) wrote:

> ... ideally, the members of the Board of Directors of the Planning Agency would be persons highly respected by the entire community, of great professional competence, be able to put forth almost 100 percent effort in the agency's work, and be completely nonpartisan. Such persons never exist (p. 366).

It is also desirable that the commission be composed of a sampling of persons representing the various classes and minorities within the client community. In actual practice, most planning commissions might not have in their ranks members representing the gamut of the society, but usually they do have representatives of the leadership.

REGIONAL PLANNING AGENCIES

Regional planning agencies (RPAs) usually serve several counties and/or cities and towns (council of governments). Their central body (usually called a commission) is composed of political and lay representatives of the client areas. Ideally, each client member will have its own planning commission without a professional planning arm. The RPA has a professional planning staff which consults with and provides professional planning assistance to the local planning commission. In a similar manner to the selection process discussed above, members of local planning commissions are elected or selected to become members of the Regional Planning Commission. An executive committee is elected, consisting of about twelve people, and meets regularly (once a month) to discuss and arrive at decisions on current planning issues. Essentially, the major service of RPAs is to provide professional planning assistance to many communities which singularly could not maintain a professional planning staff.

PLANNING COMMISSION FUNCTION

Bolan (1967) outlined planning commission function when he wrote:

> ... the planning commission (with its professional staff) is an advisory body which assists government in formulating policy. Its view is comprehensive in

that no aspect of community development is assumed to be beyond its responsibility. It is also comprehensive in the sense that the planning commission is the guardian of the whole public interest, rather than any particular special interest (p. 234).

The commission acts in an advisory capacity to the political group, and usually serves at the will of the political group. It may be charged with the responsibility of evaluating and accepting or rejecting zoning appeals or building permits, or have similar duties. In any case, the political group has the final word, and may override planning commission recommendations.

Regional planning commissions usually serve as an A-95 clearinghouse. The A-95 review process is from the Bureau of the Budget (BOB) and consists of a regional planning review of particular projects to be constructed and/or proposed activities to be initiated. Through the review process, it is determined whether these projects and/or activities are in conflict with existing plans. Usually, these projects or activities are proposals to apply for and use federal funds, and the nonapproval of an A-95 review by the RPA could mean a rejection from the federal funding agency. In this manner, the A-95 review process gives the regional planning agency a certain amount of power to wield in supervising the proper carrying out of regional plans.

Other, equally important functions provided by the larger planning commissions and the RPAs may be divided into three broad categories: (1) technical assistance, (2) application for funding and funding expedition for the client community(ies), and interrelated to these two, (3) zoning ordinance and comprehensive plan development. Technical assistance is provided by the planning staff and may consist of knowledge of a variety of fields, from sewerage problems to information on geriatric problems. This is a community service in which the planners share their expertise in consulting the community and public organizations for the organization and improvement of community service functions.

Many planners keep abreast of the latest application procedures for federal funding and the availability of federal funds, and if a community wishes to purchase a fire truck, establish a paramedic squad, build a new recreational facility or construct a new sewerage system, etc., applications may be completed by the planner(s) in cooperation with representatives from the locality. Millions upon millions of federal dollars have been brought into communities in this manner, most of which would not have been possible if not assisted by planners. Funding expedition is a very valuable service performed by planning agencies.

The third major function, that of technical assistance on zoning ordinances and comprehensive plans, is of critical importance. At this point, many communities lose out on federal funding because they have not satisfactorily completed their "homework." Many applications for federal funds require that a community have a zoning ordinance and/or a comprehensive plan.[52] These requirement(s) are appropriate in that the federal government does not wish to commit monies to a community which maintains no logical scheme of development, or in short, where the citizens have failed to make a commitment to orderly development. Because many communities in the past have been reluctant to adopt plans and zoning ordinances, they have been automatically overlooked from consideration in federal funding programs.

THE PROFESSIONAL PLANNING STAFF

As sizes of communities and planning commissions increase, so does the professional planning staff. In small communities, the planner may be a part-time building inspector, zoning ordinance administrator, or a semiprofessional of similar stature. In general, the planner tends to be a "jack-of-all-trades." However, in larger communities, a large professional planning staff may be maintained, and the planner becomes a specialist occupying one's time developing a specific program, for example, minority housing.

The planning staff may be divided into the administrative section and the planning section. The administrative section consists of the Director of Planning, Deputy Director(s), and the secretarial staff. The Planning Director or Executive Director, as suggested by Benckert (1973), should have about five years of experience, have a degree in urban planning or a related field, and ideally, be a professional engineer. The writer has found that, in practice, many planning directors do not have a degree in planning, although many do have a degree in a related field. Some planning directors are retired senior-ranked military officers, some are professors in history or economics and some do have degrees in planning. The only common denominator seems to be the fact that the planning commission agreed on the director's appointment.

By design, many planning director positions are quite secure. This is only just, in that if this person had to worry about his job whenever he or she stood up against an issue, then soon he or she would become ineffectual, in essence a "yes man." However, in reality this security has its good and its bad points, in that there are no checks against bad performance by the director. For example, if a director does go sour,[53] to remove him or her might take a majority vote by the planning commission, which in reality may be quite difficult to obtain. For one reason, many planning commission meetings do not have enough members present to achieve such a vote. Many planning commissions are di-

---

[52]See the American Institute of Planners (1975) for discussions of comprehensive plans.

[53]Planning directors have frustrating positions, in that they recommend quite just, fair and well-thought-out proposals, and planning commissions and political groups tend to reject them quite easily.

vided in vote because half of the commission members may dislike the other half. In this case, one half will vote against whatever the other half recommends, and the vote will be compromised. Problems similar to these develop, and the planning director usually need not worry about the security of the position.

THE PLANNER

Ashworth (1973) gives a definition of a planner as:

> . . . (planner) is a widely used (and misused) word. It has no specific legal definition. There are economic planners, traffic planners, regional planners, and many more. They are not even necessarily professionally or technically trained. "Planner" in the local newspaper may refer to ministry official, local government planning officer, member of parliament, or local councilor. . . . To be a little more specific, however, within the context of land planning the planner is generally taken to be the man who has to cover the whole range of social, economic and physical factors which make up the context of life, and who tries to make preparations for its future demands (p. 74).

Today, many professional planners do have degrees in planning or a closely related field. However, by the nature of their work, a degree in planning may not be required. For example, as has been discussed above, many planning agencies, especially RPAs, provide technical assistance to their client communities. These planners continuously keep abreast of a particular part of the federal funding theater, and although they may be called "planners," a degree in criminal justice, geriatrics or civil engineering may be the appropriate requirement for fulfilling such a position. Usually, a masters degree with experience is required for the position of planner; junior planners may require a bachelors or a masters degree.

The planner's position is by no means as secure as that of the planning director. The planner may be hired by the planning commission in conjunction with the planning director, but in most cases he or she is hired directly by the director, who may fire the planner at any time. If the person is a professional planner and wishes to maintain a planning career, a firing usually initiates an exhaustive job search. If the planning director has soured, planners on his or her staff suffer the consequences of working for a tyrant. The director, having lost many battles against the planning commission or various political groups, finds the staff a good "whipping boy" for his frustrations, as Levin (1976) observes when he calls a planning director a "bully boy," and writes:

> . . . some planners are shoved around by supervisors they heartily dislike and fear; others continue on in a work environment they detest (p. 8).

If the planner sees eye to eye with the planning director, they can find solace in their rejections and develop new, effective strategies. If he or she is at odds with the director, the planner faces rejection from both sides. All in all, the job of planner is difficult. If he or she is a conscientious, good-willed person who makes logical and prudent recommendations to a political group who have other concerns in mind and reject these proposals, the planner becomes frustrated, and may find that there is no way to vent those frustrations. Eventually, more often than not, the planner may also go sour, and become a bureaucrat. As Levin (1976) writes:

> . . . goofing off on the job, occupying space, drawing a salary but doing nothing productive. Some planners may comfort themselves with the notion that they are learning their trade. But much of this is self-delusion and comforting rationalization. Like the rest of the army of useless public and private bureaucrats, far too many planners are engaging in this most costly game (p. 8).

6.2 The Plan and the Planning Process

THE PLAN

Black (1975) defined a comprehensive plan:

> . . . (a plan) is an official public document adopted by a local government as a policy guide to decisions about the physical development of the community (p. 1).

The comprehensive plan (in some cases called a "master plan") should predate the community zoning ordinances, official maps and subdivision regulations. The latter are specific and detailed pieces of legislation which are intended to carry out the general recommendations of the comprehensive plan.

The purpose of the comprehensive plan is to guide decisions regarding the provision of public services for existing and future structures and other uses of the land.

The essential characteristics of the plan are that it is comprehensive, general and long-range. "Comprehensive" means that the plan encompasses all geographical parts of the community and all functional elements which bear on physical development. "General" means that the plan summarizes policies and proposals and does not indicate specific locations or detailed regulations. "Long-range" means that the plan looks beyond the foreground of pressing current issues, to a desirable possible future twenty to thirty years hence (Lazaro, 1977a).

Black (1975) pointed out the need for a comprehensive plan when he wrote:

> . . . the local government is inescapably involved in questions of physical development . . . the local government—and particularly the legislative body made up of lay citizens—needs some technical guidance in making these physical development decisions. This guidance can be provided by professional city planners . . . on the basis of expediency of ad hoc "quickie" studies, then there is no

guarantee that next month's decision will not negate the one made today. The local government needs an instrument which establishes long-range, general policies for the physical development of the community in a coordinated, unified manner, and which can be continually referred to in deciding upon the development issues which come up every week. The comprehensive plan is such an instrument (p. 3).

The document itself may come in various sizes and shapes and may have many outlines. We will not attempt to describe these features here. The planning process which follows lists the fundamental phases of plan development.

THE PLANNING PROCESS

Priscoli (1975) defined planning as ". . . a process of developing programs of action to meet goals" (p. 1234). Litchfield, Kettle and Whitbread (1975) defined a planning process as:

> . . . a course of activity that is intended to heighten understanding of the nature of problems requiring examination of the alternative possible solutions that exist, and of the relative merits of these alternatives (p. 18).

In concept, this process may be modeled quite easily. The author (Lazaro, 1978) modified a schematic model of the planning process by Malone (1973):

> *Phase I—Goals*: Determination of clear-cut goals by the planner working with the local planning advisory group.
>
> *Phase II—Study*: Gathering of pertinent data and the writing of a description of the planning area problems to develop an understanding of the various needs.
>
> *Phase III—Analysis and Synthesis*: Working in consultation with the local planning commission or advisory group, the planner defines the needs of the locality and forms the basis for his recommendations.
>
> *Phase IV—Recommendations*: The arguments and needs having been developed are now placed in the form of recommendations for future courses of action.
>
> *Phase V—Adoption*: The recommendations are presented to the political body for acceptance and implementation.
>
> *Phase VI—Implementation*: The political body resolves that a certain course of action be enacted to implement the recommendations.
>
> *Phase VII—Update*: A planning area is not a static entity, but varies with time, and plans must be updated periodically to reflect the changing nature of the future needs (p. 2).

The planning process is the most crucial part of the planning effort. To illustrate the planning process, a case study will now be presented. Although the case study is a brief technical report, and is not as elaborate as a formal plan,[54] it was selected because of its simplicity and its ability to lend itself to the planning process outlined above (Lazaro, 1978).

[54]For a good outline of the comprehensive planning process, see the American Institute of Planners (1975).

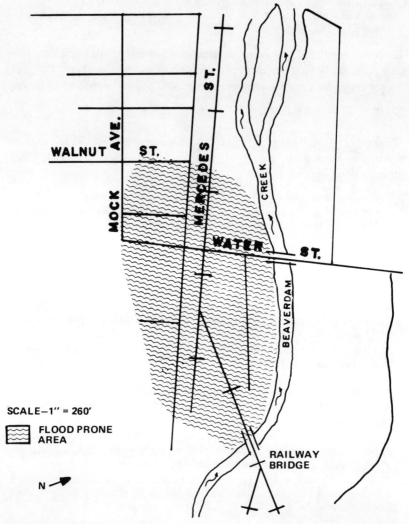

Figure 6.2. Flood-prone area [*Source*: Lazaro, *J. Env. Systems*, "The Planning–Political Interface: A Lesson in Plan Adoption," 8(1):1–11 (1978), © 1978, Baywood Publishing Co., Inc.].

CASE STUDY: BEAVERDAM CREEK FLOOD PREVENTION

The case study was conducted in a small, rural community in southwest Virginia, a pleasant little village with a population of about 1200. The community is nestled in a narrow, mountainous river valley surrounded by steep knobs. Long ago, farmers cultivated the rich flood plain, modifying its surface structure. When a railroad spur was extended to the town, growth was stimulated. Since there was little vacant land left for development, any former agricultural areas were occupied by housing, and the capability of floods to cause damage increased.

Phase I—Goal: The Prevention of Flooding: A few years ago, the author (then a planner for a regional planning agency) was requested by the town council to do a study of flood problems along Beaverdam Creek and to make flood prevention recommendations.

Phase II—Study: Approximately twenty structures are contained in the flood-prone area. The most severely affected sections—the land along Mercedes Street, Mock Avenue and Water Street—are depicted in Figure 6.2. Significant damage occurred in the floods of 1901, 1940, 1955 and 1957 (Tennessee Valley Authority, 1957). However, minor floods occur much more often, affecting parts of the total area and initiating the request for this study.

Phase III—Analysis and Synthesis: The analysis was based on studies done by Wolman (1967), the author (Lazaro, 1976), and on discussion by Maddock (1976). Several residents of the flood-prone area stated that the flood waters always followed a certain path; flooding began at the lowest point next to the dike and the railway bridge.

Upon further investigation, it was observed that the surface of the flood plain along the creek has been extensively altered by man. Figure 6.3 reveals that on the western side of the creek, a dike reduces the cross-sectional area of the stream. The eastern side of the creek has a higher (less modified) flood plain. Having a smaller flood plain to buffer stormflows, the stream's ability to handle overbank flows is seriously reduced at this point.

In addition, it was found that within the stream channel, flooding conditions are further aggravated by manmade constrictions: a railway bridge, a stone masonry wall on the eastern bank and a shoring made of boulders on the western bank. The bridge piling is an obstacle to storm water flow because debris may be held suspended by the piling. Also, the bridge does not provide sufficient clearance for high flood waters, since it is only eight feet above the streambed. When storm waters rise above this height, the bridge acts as a partial dam, increasing the impoundment of water and augmenting the volume of water spilling over the dike.

Moreover, once the water is impounded, stream velocity decreases and sedimentation follows. A massive sandbar upstream of the bridge gave evidence of this. The sandbar also serves to reduce the cross-sectional area of the

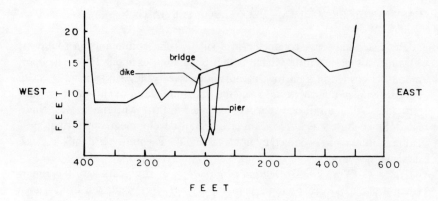

Figure 6.3. Beaverdam Creek stream cross section at Railway Bridge. If the dike were not present, the stream could overflow onto the flood plain area which lies about 100 ft to the west of the dike before encountering a natural obstacle. However, housing has encroached upon this area, and when the creek overflows, damage results [*Source*: Lazaro, *J. Env. Systems*, "The Planning-Political Interface: A Lesson in Plan Adoption," 8(1):1-11 (1978), © 1978, Baywood Publishing Co., Inc.].

stream, which increases the flood potential in successive high water conditions. Thus, the flood potential damage capability becomes progressively more serious. In a natural stream, it has been estimated that bankfull flow occurs once every 1.5 yr (Leopold et al., 1964). Within this reach, the frequency has been considerably increased. Because the stream's western flood plain has been encroached upon by residences, when an overbank flow is produced, damage results.

Phase IV—Recommendations: It has been suggested by Debo (1976) that many local political groups do not have the expertise to properly interpret technical studies. Since this was a rather technical report, and was to be presented to a lay group (the town council), utmost care had to be taken to communicate an accurate understanding of the flood preventive measures. The author has reported the logic of communication in more detail elsewhere (Lazaro, 1977b).

Presentation of the recommendations to the town council followed procedures similar to those proposed by Estrin and Monahan (1975). Visual aids were employed; in this case, 35 mm slides. A brief report showing cross sections of the creek, listing all the recommendations and explaining the causes of flooding, was given to each council member. Slides of Beaverdam Creek a mile upstream of the railway bridge in a forest setting were shown first. Their purpose was to illustrate an unmodified natural stream (i.e., a stream in equilibrium, or in balance with its environment). Then, contrasting slides of the stream in the immediate area of the railway bridge were shown to demonstrate a stream in disequilibrium (i.e., distorted by man-induced land surface changes). Each recommendation was presented with the aid of slides and explicitly described to the council members.

The following recommendations with explanations were proposed in order of priority:

(1) *Removal of the railway bridge*. Since the bridge has not been used in several years (the spur terminates a few hundred feet past the bridge), it was recommended that it be removed and the accumulated sediment cleared. In this manner, the cross-sectional area would more closely approximate the larger, natural one, and the chances of flooding would be greatly reduced (estimated cost to the town: $2,500).

(2) *Tunnelling through the western bank and sediment removal*. A tunnel under the railway track on the western bank attended by sediment clearance would increase the volume of the streambed and alleviate flooding at this point. In order for this alternative to remain effective, sediment removal would have to be conducted periodically, which would represent a recurring expense to the town (estimated cost to the town: greater than $2,500).

(3) *Channelization*. Channelization involves a more extensive and costly restructuring of the stream to widen and deepen the channel. A preliminary examination indicated that channelization may be prohibitively expensive because of the proximity of residential structures to the stream and the fact that the streambed appears to be bedrock (estimated cost to town: less than $50,000).

(4) *Purchase and conversion of the flood prone area to other uses*. It must be recognized that the existing topography, i.e., the low western flood plain blocked by the relatively high dike, will continue to make the Mercedes Street area of the town susceptible to flooding. This suggests that there is no "cure" for flooding problems in this area. All of the recommendations are alleviative measures which will considerably reduce the probability of flooding, but will not bring about its complete elimination. Therefore, in the long run, the adoption of any one or all of the above recommendations may prove to be more expensive and less effective than for the town to purchase the encroached flood plain areas, remove the structures, and convert the land to other, less damage-prone usages, e.g., park land (estimated cost to town: less than $50,000).

Phase V–Adoption: The town council did not choose to implement any of the recommendations. Approximately a month later, they engaged a local contractor to use a bulldozer to clear the streambed of loose stones and smaller sediment (Guy, 1976).

The author at first could not understand why this action was taken, since it represented a temporary, symptomatic cure. After a few high water flows, the stream would again become flood prone and again require corrective maintenance.

Upon further investigation, the author determined that the primary cause for the action undertaken was as follows. Voters from the flood-prone area believed that by clearing the stream of debris, flooding would cease. Because this was September of a local election year, the town council was sensitively tuned to voter desires. Proposal (1), removal of the railway bridge, would require more time to accomplish, even though it would have most likely cost the same as the action taken. In the pursuit of reelection, the bridge was left standing and a symptomatic approach was undertaken.

THE PLANNER AS A POLITICAL INSTRUMENT

The preceding case study demonstrates one of the author's experiences in a planning process, and it briefly illustrates the pervasiveness of the planning–political interface. The case study concerned about a month of actual time, and therefore all the various phases of the planning process could not properly be developed. In other words, all of the persuasive tools that a planner could employ were not allowed to develop successfully into influential arguments to modify political decision making. One very effective tool is citizen participation, which will be discussed in greater detail later.

In the previous study, the planner was used as an instrument to bring about what the political group most likely had intended to accomplish. In other words, the course of action was already decided prior to the planning study. Whatever the planner might recommend was not an issue to be considered.

The author had these experiences in several different studies and observed other planners who were similarly exploited. This case study is not unique in this respect, and this type of exploitation occurs quite often. It must be pointed out here that, on the other hand, a few successful exploitive experiences (i.e., the political group uses the planner to accomplish a goal which coincidentally enhances the planner's career) have taken planners up the ladder to success. However, in either a positive or negative case, a true planning process is not effectively developed, and planning is definitely not an integral part of the final decision.

PLANNING AND POLITICAL DECISION MAKING

This leaves us with the question, "How can a planner affect political decision making?" What are these tools which may enhance the planning process?" By developing an understanding of the political decision-making process, an effective strategy may be formalized.

Catanese (1974) discusses an approach toward modeling the political process taken by Easton (1965). Figure 6.4 shows that the input to the political system conversion mechanism consists of two major political variables: supports and demands. Special interest groups and key individuals generate the demands.

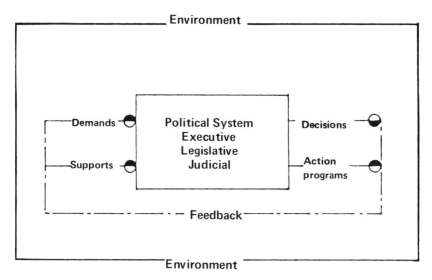

Figure 6.4. The political process [reprinted from *Planners and Local Politics: Impossible Dreams* by Anthony James Catanese, Sage Library of Social Research, Volume 7, © 1974, p. 38 by permission of the publisher, Sage Publications, Inc. (Beverly Hills/London)].

Planning agencies and other institutionalized agencies may also generate demands. Catanese points out that for any demand or set of demands, there is a corresponding set of support which may be positive, negative or neutral, in varying degrees.

The political system receives these demands and supports, weighs them and intuitively will act or take no action. Decisions are the responses to the supports and demands. Action programs are commitments, usually through the allocation of resources to meet the decisions. The feedback loop in the political process is necessary in order to provide both stability and change. As Catanese writes:

> . . . a political system with no feedback loop can produce decisions and programs that are so obnoxious to the society that it leads to overthrow and replacement of the system (p. 39).

In the planning process (illustrated above), the political decision-making group made an intuitive judgment of the inputs of voter demands and supports and implemented a decision which was aimed to assure reelection. The planner's recommendations, on the other hand, were largely objective. The planner's experience and training in his field of expertise allowed him to make judgments on a scientifically based course of action.

Demand and support inputs are greatly affected by the public and, because politicians are tuned to the public will, public participation is a crucial part of the planning process. Willeke (1974) points out the need for public participation in planning when he writes:

> ... planning is fundamentally a social, rather than a mathematical/logical process. It is a process of ascertaining needs, goals, desires, and objectives; diagnosing problems; formulating alternatives; and testing, proposing and implementing. In each of these stages, interaction with people in social situations is necessary to complete the stage satisfactorily (p. 79).

Fundamental to public participation is the planner's establishment of legitimacy and trust. As Willeke states:

> ... the greater the legitimacy of the planner and the more he is trusted, the more likely it is that the public will be willing to enter into talks with him, and the more likely it is that the resulting plan will be favorably received by the public (p. 80).

Willeke schematically models the communicative process and discusses the need for two-way communication. Public participation comes in many forms, and it is essentially based on democratic concepts. Very simply described, it is the citizenry who must live with the adopted plan. If the people are included in the planning process, their voices will have been heard and will have had an influence on the various phases and ultimately the final work. Furthermore, it is the people who must obey and in essence enforce the recommendations of the plan; enforcement agencies cannot possibly be large enough to supervise thoroughly the carrying out of the plan's directives. If the people have input into the planning process and accept the final plan, the democratic process is at work and enforcement is a minor, if not totally unimportant, issue.[55] In short, public participation, if handled properly, allows two-way communication and the resolution of conflicts (Priscoli, 1975); through communication, an appreciation of each person's role is established, planner trust and legitimacy are developed and the final work plan becomes a truly applicable legal instrument.

The preceding paragraph describes the ideal bases for public participation. In reality, there are authors who would question the public's ability to participate as Dye and Zeigler (1972) suggest that large numbers of the electorate are uninformed politically, and generally unsupportive of democratic values. Levin (1976) questions public representation when he summarizes:

> ... as a practical matter, the public has limited time and technical expertise at its disposal. Attempts to secure a wide spectrum of citizen participants usually fail. As a result, citizen participation generally takes on a representative character: trusted volunteers or paid experts represent the interests of the hitherto neglected segments of the population. In contrast, effective mass participation

---

[55]Public Law 92-500 will be discussed below; enforcement is a critical part of the law.

usually occurs only as a negative response to a specific proposal, such as the construction of a highway through a residential neighborhood (p. 9).

Adequate public representation and participation are serious problems in planning. Apathy and ignorance on behalf of the common citizen is difficult if not totally impossible to overcome. Without active, concerned citizens, the planner's functional capabilities are greatly limited. However, one means that may be employed to enhance the planning effort is to establish close working relationships with community influentials.[56] (The author points this out in Lazaro, 1977b.)

6.3 Urban Water Resources Planning

The supply of drinking water and the removal of dirty water are regular expenses which make up a large part of the budget of an urbanized area. The planning of such systems, urban water resources planning, is an integral part of community comprehensive planning. Urban water resources planning en compasses two broad fields: surface water supply and distribution, and storm, sanitary and combined sewerage. This book has discussed urban water quantity and water quality runoff problems, and thus has not considered urban water supply and distribution. Accordingly, this section will present a brief discussion on a current issue in urban water resources planning which has a direct bearing on sewerage as well as nonpoint source pollution.

WATER QUALITY MANAGEMENT PLANNING

Water quality management planning gained considerable momentum in response to the environmental movements of the late 1960s and early 1970s. Much of this interest has been further motivated by Public Law 92-500, the Federal Water Pollution Control Act, as amended in 1972. In Section 101(a), 33 U.S.C., Section 1251(a), is found the general goal of this legislation, which is: ". . . the restoration and maintenance of the chemical, physical and biological integrity of the Nation's waters."

The nation's waters were to be fishable and swimmable by 1983. This was to be achieved in several phases. The original focus of Public Law 92-500 was to clean up the easily visible and offensive point source discharges. Under Section 201, millions of dollars were allotted to communities to update their existing sewage treatment plants and/or construct new ones which would satisfy the law mandates. This expenditure of vast amounts of money came under

---

[56]A community influential is a person in a position to exert significant force on the direction of community change (Hillegass et al., 1970).

fire by several scholars, perhaps the leader of which is General Whipple. As Whipple et al. (1974) conclude:

> ... in many developing urban and suburban areas, once the stage of secondary treatment of recorded wastes (point sources) is arrived at, the unrecorded pollution sources (nonpoint sources) will account for more than half of the pollution in the streams (p. 884).

In general, the argument upheld by these authors was that in the haste of trying to get the program moving, improper consideration was given to determining the most cost-effective measures and approaches which would effectively reduce water pollutant discharges. In his book, Whipple (1977) comprehensively develops this argument and brings forward some very good points. He discusses the Environmental Protection Agency's (the administrative agency for Public Law 92-500) planning approach as being illogical. Whipple writes:

> ... the act (Public Law 92-500) calls for basinwide planning (Section 303e), areawide planning (Section 208), and facilities planning (Section 201). The EPA has chosen to carry out the planning in a somewhat illogical order. To get work under way rapidly, major emphasis and funding was put first in preparing facilities plans. States were given grants for basinwide planning, but the sums alloted [sic] and trained personnel available were totally inadequate to do the job.

Table 6.1 lists the National Commission on Water Quality estimated costs for implementing Public Law 92-500. The total is $511.5 billion. This is a fantastic amount, making the implementation of the law mandates economically questionable, if not totally impossible.

Section 208 authorized the development of areawide clean water plans for

Table 6.1. Estimated costs for implementing PL 92-500.

Costs	Billion $
Costs of achieving 1977 treatment goals for publicly owned plants, federal outlay only (75%) (For 1983 goals practically the same cost)	118.5
Nonfederal costs of achieving 1977 goals for publicly owned plants would presumably be 1/3 of federal cost or	39.5
Correction of combined sewer overflows	79.6
Treatment of urban runoff only to the extent of suspended solids removal and disinfection	199.0
Capital costs for industry to meet 1977 goals	44.3
Added costs for industry to meet 1983 goals	30.6
Total	511.5

Source: Whipple (1977). Reprinted by permission of the publisher, from William Whipple, Jr., *Planning of Water Quality Systems*, p. 146.

areas with serious water pollution problems as a result of urbanization or industrialization. Davis (1976) reports that:

> . . . Section 208 recognized that urban water pollution has become so serious that waste treatment technology alone could not provide a complete solution. Therefore, areawide planning incorporates wider concerns such as land use, growth management, air quality, nonpoint source pollution, and solid waste disposal in an areas' management plan. But the major component is *management*. That is, the final plan must recommend a management structure capable of implementing its component technical solutions (p. 38).

Whipple (1977) lists seven overall planning features of Section 208 (see Figure 6.5):

(1) Identification of problems
(2) Identification of constraints and priorities
(3) Identification of possible solutions to problems
(4) Development of alternative plans
(5) Analysis of alternative plans
(6) Selection of an areawide plan
(7) Periodic updating of the plan

This list, as well as Figure 6.5, indicate that the EPA planning strategy is excellent in theory. All the major elements of sound planning have been incorporated in addition to citizen participation, which is an integral part of the entire planning effort. However, in actual practice hasty data collection programs and poor coordination between the planning accomplished under the three sections (303, 201 and 208) left much to be desired. The practical relationships between 303, 208 and 201 studies in a region of Northeast New Jersey have been discussed by Dobrowolski and Grillo (1977). The authors report:

> . . . preliminary evaluations show that detailed 201 work can affect the conclusions of 303 and 208 studies, and that a wider (environmental-social as well as economic) interpretation of cost-effectiveness can demand re-examination of prior assumptions and decisions, a task not typically part of 208-303 work (p. 455).

They suggest that the U.S. Congress establish a mechanism to see that funding be available for 201 agencies for participation in 303 and 208 studies. They emphasize that studies should be practical and consider details where appropriate and make recommendations consistent with the level of the study performed.

Enforcement of Public Law 92-500 is perhaps the most serious failing of the act. Assuming that the 201, 303 and 208 studies are completed and implemented adequately, what force will maintain the integrity of the enhanced quality of U.S. water? Whipple outlines the problem:

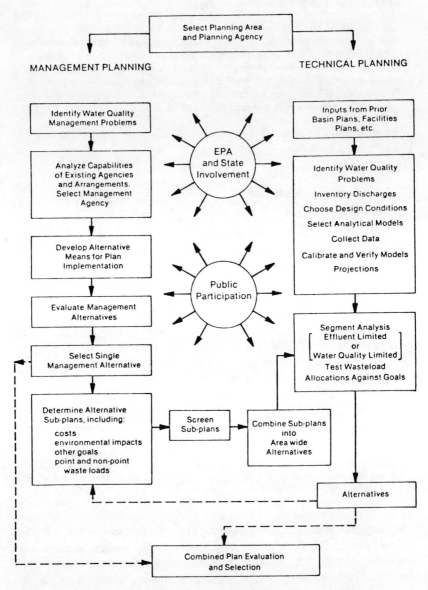

Figure 6.5. Section 208 planning process [*Source*: Whipple (1977)].

> ... under the present situation, the federal EPA and the states have virtually absolute authority over point source discharges, through financing of treatment plant construction and through the system of NPDES permits[57] described above. However, authority and procedures are unclear as regards local land use controls and nonpoint sources. The local agencies performing Section 208 planning were chosen in compliance with EPA criteria; in particular, they were required to have a land use plan capability. However, they were not required to have powers to enforce land use controls, or any function or expertise whatsoever in the water quality field (p. 149).

Thus, a strategy to meet and overcome the enforcement problem is indeed perplexing. Say and Dines (1977) have examined and presented a logical and effective approach to solving this problem as well as other water resources planning problems. The first step (which is also the major disadvantage of this method) is to initiate preventive approaches instead of corrective ones. Because many communities are dealing with water resource problems after the fact, preventive approaches are not possible in many cases, and many communities are automatically eliminated. A major impediment to the application of a preventive approach in communities is, as Say and Dines write:

> ... water resources protection programs may require local elected officials to make and enforce decisions which reduce windfall speculative profits of constituents, who vote, in order to promote future savings to the community at large, who may have no way of voting (p. 914).

The second step is to choose small watersheds (creeksheds) as the organizing unit. If these areas are improperly managed during urbanization, they may require extensive and expensive public works programs. Creeksheds are closest to the people in a political and geographic sense, and can be affected most directly by the actions of local units of government. Creekshed residents are part of the multiple-phase information feedback process, and provide a proper citizen participatory mechanism. Residents may engage in the planning process at a local level. Large river basin planning lacks such intimacy, and people tend to feel alienated. They may feel that the boom is being lowered on them, and they have no choice in the final outcome.

In contrast to Public Law 92-500 and the EPA's top-down approach, this "grass-roots" approach, ideally, would be a viable means of structuring a comprehensive water quality management planning strategy. However, it too has its drawbacks. Practical consideration must be given to apathy as a limitation to the proper application of both approaches. Planning commissions and planners must continually try to motivate public interest in planning issues. This is not an easy task. A well-coordinated public participation program may do much to alleviate apathy problems.

---

[57] NPDES permits are EPA permits specifying conditions of compliance with federal law as to allowable wastes to be discharged by point source discharges.

As has been pointed out in this discussion, Public Law 92-500 has had its problems. However, they are not totally irresolvable, and even though they have met with much dispute, the law has been flexible enough to allow some latitude. It is continuously being modified to consider new situations. The goal of Public Law 92-500 is an extremely bold one, considered by many to be too ambitious. The goal of Public Law 92-500 of fishable, swimmable waters by 1983 was not realized. However, the day will not be too far off when the primary goal will be accomplished. One must realize when one looks into the future that the future is achieved one day at a time. What must be attempted now is a working, day-to-day plan. To go from the 1972 water quality deterioration problems to fishable, swimmable waters in eleven years was too ambitious. But, at the same time, such an action must be commended, for it represents an invigorating breath of fresh air in an attempt to solve an extremely difficult multidisciplinary problem.

6.4 Summary

> . . . planning is that activity that concerns itself with proposals for the future, with the evaluation of alternative proposals, and with the methods by which these processes may be achieved (Simon et al., 1950, pp. 423–424).

The growth of population and industry related to the industrial revolution made planning a necessity. Planning today is practiced by a majority of the more than 35,000 township and municipal governments and the 3049 county governments in the United States.

A planning commission is a group of local citizens who are usually appointed by the mayor or the town council with authority from state enabling legislation. These people serve without pay as a civic duty. Ideally, they are representatives of the classes and minorities of the client locale. The major function of a planning commission is to advise the political decision-making group on policy. Regional planning commissions are made up of elected or selected members from local planning commissions.

As localities become larger and their planning needs grow, they may not only have a lay planning commission, but also they may hire a professional planning staff to advise the commission and in turn be advised by the commission. The planning staff, depending on the size and needs of the client community, may supply expertise in many fields, and greatly enhance community services. The regional planning agency usually supplies planning functions for several counties and/or communities which individually could not hire and maintain such planning expertise.

A comprehensive plan provides a long-term guide for the community leadership. It promotes the making of decisions which are compatible with orderly community growth and development. Seven steps in the planning process are outlined and a case study of a planning process is illustrated. The differences between a planner's objectives and political objectives/decision making is demonstrated. Briefly, the politician responds to intuitive feelings of his or her constituency's needs, whereas the planner is largely objective in the courses he or she recommends.

Urban water resources planning is a vital part of urban comprehensive planning, making up a large part of the urban budget. Water quality management planning is discussed relative to Public Law 92-500, the Federal Water Pollution Control Act, as amended in 1972. This law has met with controversy in that the approaches taken to achieve the restoration and maintenance of the chemical, physical and biological integrity of the nation's waters have been illogical. Say and Dines (1977) present a very logical, grass-roots planning strategy which may greatly facilitate enforcement. The essential ingredients are a preventive instead of a corrective approach and work with creekshed (small watershed) units. This planning strategy has its drawbacks in that most urban areas by necessity must practice corrective water resources planning, and general public apathy must be overcome. The latter, in general, is a chronic problem in comprehensive planning, and it is a cross that many planners must bear.

6.5 Exercises

Considering an urban area with which you are very familiar:

1 Select what you feel is the most critical urban hydrologic problem (i.e., increased erosion, inadequate sewerage system, pollution of the stream, etc.):
   a) What do you think would be the best planning strategy to solve this problem?
   b) In a block diagram form, model your strategy.
   c) Defend your strategy.
   d) Would you collect data? What kind of data?
   e) What methods would you use to illustrate and prove the need for the solution to this problem?
   f) Which part of the planning process would be the most difficult for you to complete? Why?

References

AMERICAN INSTITUTE OF PLANNERS. "The Comprehensive Planning Process: Several Views," Washington, DC:The American Institute of Planners, 172 pp. (1975).

ASHWORTH, G. *Encyclopaedia of Planning*. London:Barrie and Jenkins, 120 pp. (1973).

BENCKERT, K. W. "Funding and Supporting Planning," in *Handbook on Urban Planning*. W. H. Claire, ed. New York:Van Nostrand Reinhold Company, pp. 363–372 (1973).

BLACK, A. "The Comprehensive Plan," in *The Comprehensive Planning Process: Several Views*. Washington, DC:American Institute of Planners, pp. 1–30 (1975).

BOLAN, R. S. "Emerging Views of Planning," *J. Am. Inst. Planners*, 33(4):233–245 (1967).

CATANESE, A. J. *Planners and Local Politics: Impossible Dreams*. Vol. 7 in Sage Library of Social Research. London:Sage Publications, 189 pp. (1974).

CLAIRE, W. H., ed. *Handbook on Urban Planning*. New York:Van Nostrand Reinhold Company, 393 pp. (1973).

DAVIS, R. "208–A Report," *Practicing Planner*, 6(4):38–45 (1976).

DEBO, T. N. "The Complete Drainage Program: More Than an Ordinance," *Water Resources Bull.*, 12(1):109–121 (1976).

DOBROWOLSKI, F. and L. Grillo. "Experience with the 303-208-201 Study Relationships," *Water Resources Bull.*, 13(3):455–460 (1977).

DYE, T. R. and L. H. Zeigler. *The Irony of Democracy*. Belmont, CA:Wadsworth Publishing Co., Inc., 396 pp. (1972).

EASTON, D. A. *A Systems Analysis of Political Life*. New York:John Wiley and Sons, Inc., 507 pp. (1965).

ESPEY, W. H., JR. and D. E. Winslow. "Urban Flood Requency [sic, Frequency] Characteristics," *J. Hyd. Div.*, ASCE, 100(HY2):279–293 (1974).

ESTRIN, H. A. and E. J. Monahan. "Effective Oral Presentation of Scientific and Technical Information," *J. Tech. Writing and Comm.*, 5(3):187–197 (1975).

EWING, D. W. *The Human Side of Planning*. London:Collier-Macmillan Ltd., 216 pp. (1969).

GUY, M. "Virginia's Circuit-Riding Administrator," *Appalachia*, pp. 32–36 (1976).

HILLEGASS, T. J. et al. "Community Decision Structure and Urban Planning Process," *J. Urban Planning & Dev. Div.*, ASCE, 96(UP1):17–22 (1970).

LAZARO, T. R. "Adapted Techniques for Urban Stream Structure Analysis," *J. Environ. Systems*, 6(4):321–328 (1976).

LAZARO, T. R. "Comprehensive Plan for the Town of Independence, Virginia," Mount Rogers Planning District Commission, Marion, VA, 75 pp. (1977a).

LAZARO, T. R. "Effective Communication of Technical Information to a Nontechnical Group," *J. Tech. Writing and Comm.*, 7(4):295–301 (1977b).

LAZARO, T. R. "The Planning–Political Interface: A Lesson in Plan Adoption," *J. Env. Systems*, 8(1):1–11 (1978).

LEOPOLD, L. B. et al. *Fluvial Processes in Geomorphology*. San Francisco:W. H. Freeman & Company, Publishers, 522 pp. (1964).

LEVIN, M. R. "The Conscience of the Planner," *Planning*, January 8, 1976, pp. 8–10 (1976).

LEWIS, H. M. *Planning and the Modern City*. New York:John Wiley and Sons, Inc., 224 pp. (1949).

LICHFIELD, N. et al. *Evaluation in the Planning Process*. New York:Pergamon Press, Inc., 326 pp. (1975).

MADDOCK, T. JR. "A Primer on Flood Plains Dynamics," *J. Soil Water Cons.*, March/April, 1976, pp. 44–47 (1976).

MALONE, W. A. "Planning Practice and Techniques," *J. Soil Water Cons.*, January/February, 1973, pp. 21–24 (1973).

PRISCOLI, J. D. "Citizen Advisory Groups and Conflict Resolution in Regional Water Resources Planning," *Water Resources Bull.*, 11(6):1233–1242 (1975).

SAY, E. W. and A. J. Dines. "New Requirements for Local Units of Government in Water Resources Planning: Insights for Implementation from Recent Water Resources Planning Research," *Water Resources Bull.*, 13(5):907–915 (1977).

SEELEY, J. R. "What is Planning—Definition and Strategy," *J. Am. Inst. Planners*, 28(2):91–97 (1962).

SIMON, H. A. et al. *Public Administration*. New York:Alfred A. Knopf Inc., 582 pp. (1950).

TENNESSEE VALLEY AUTHORITY. "Floods Along Laurel and Beaverdam Creek" (1957).

WALKER, R. A. *The Planning Function in Urban Government*. Chicago:The University of Chicago Press, 410 pp. (1950).

WHIPPLE, W. JR. et al. "Unrecorded Pollution from Urban Runoff," *J. Water Poll. Control Fed.*, 46(5):873–885 (1974).

WHIPPLE, W. JR. *Planning of Water Quality Systems*. Lexington, MA:Books, 236 pp. (1977).

WILLEKE, G. E. "Theory and Practice of Public Participation in Planning," *J. Irr. and D. Div.*, ASCE, 100(IR1):75–86 (1974).

WOLMAN, M. G. "A Cycle of Sedimentation and Erosion in Urban River Channels," *Geografiska Ann.*, 49A:385–395 (1967).

Selected Readings

Advisory Commission on Intergovernmental Relations. "Multistate Regionalism: A Commission Report," U.S. Government Printing Office, #A-39 (1972).

Altshuler, A. A. *The City Planning Process: A Political Analysis*. Ithaca, NY:Cornell University Press, 466 pp. (1969).

Banfield, E. C., ed. *Urban Government: A Reader in Administration and Politics*. New York:The Free Press, 718 pp. (1969).

Brown, L. R. "Social Well-Being and Water Resources Planning," *Water Resources Bull.*, 12 (6):1181–1190 (1976).

Cartee, C. P. and D. C. Williams, Jr. "A Study of Coastal Pollution and Agency Interface," *Water Resources Bull.*, 14(5):1167–1175 (1978).

Cartwright, T. J. "Problems, Solutions and Strategies, A Contribution to the Theory and Practice of Planning," *J. Am. Inst. Planners*, 39(3):179–187 (1973).

Chapin, F. S. *Urban Land Use Planning*. Urbana, Illinois:University of Illinois Press, 498 pp. (1965).

Clavel, P. "Planners and Citizens Boards: Some Applications of Social Theory to the Problem of Plan Implementation," *J. Am. Inst. Planners*, 34(3):130–139 (1968).

Dee, N. et al. "An Environmental Evaluation System for Water Resources Planning," *Water Resources Res.*, 9(3):523–535 (1973).

Delleur, J. W. et al. "Interactions Between Land Use and Urban Water Resources Planning," *Water Resources Bull.*, 12(4):759–778 (1976).

DeLucia, R. et al. "A Water Quality Planning Model with Multiple Time, Pollutant and Source Capabilities," *Water Resources Res.*, 14(1):9–14 (1978).

Doerksen, H. R. and J. C. Pierce. "Citizen Influence in Water Policy Decisions: Context, Constraints and Alternatives," *Water Resources Bull.*, 11(5):953–962 (1975).

Douglas, J. L. "Role of Economics in Planning Flood Plains Land Use," *J. Hyd. Div.*, ASCE, 98 (HY6):981–992 (1972).

Dysart, B. C., III. "Water Quality Planning in the Presence of Interacting Pollutants." *J. Water Poll. Control Fed.*, 42(8):1515–1529 (1970).

Ellis, R. H. "New Considerations for Municipal Water System Planning," *Water Resources Bull.*, 14(3):542–553 (1978).

Etzold, D. J. "Benefit-Cost Analysis: An Integral Part of Environmental Decisions," *J. Env. Systems*, 3(3):253–256 (1973).

Faludi, A. *A Reader in Planning Theory*. New York:Pergamon Press Inc., 399 pp. (1973).

Faludi, A. *Planning Theory*. New York:Pergamon Press, 306 pp. (1976).

Ferrara, T. C. et al. "Public Participation in Urban Water Planning," *J. Urban Plan. & Dev. Div.*, ASCE, 97(UP2):179–190 (1971).

Finnemore, E. J. "Stormwater Pollution Control: Best Management Practices," *Journal of Environmental Engineering Division*, American Society of Civil Engineers, 108(EE5):835–851 (1982).

Fletcher, W. W. "The Land Use Implications of the Clean Air Act of 1970 and the Federal Water Pollution Control Act Amendments of 1972," U.S. Library of Congress, Congressional Research Service, NA 9000, 74-176 EP, 25 pp. (1974).

Foster, J. H. "Flood Management: Who Benefits and Who Pays," *Water Resources Bull.*, 12(5):1029–1040 (1976).

Fox, I. K. "We Can Solve Our Water Problems," *Water Resources Res.*, 2(4):617–623 (1966).

Frank, J. E. "The Renaissance in Land Use and Its Role in the Solution of Environmental Problems," *J. Env. Systems*, 3(3):171–188 (1973).

Friedmann, J. "A Response to Altshulter: Comprehensive Planning as a Process," *J. Am. Inst. Planners*, 31(3):195–197 (1965).

Griffith, J. "The Role of Social Scientists in River Basin Planning: A Critique," *J. Env. Systems*, 3(2):131–152 (1973).

Guy, H. P. "Sediment Control Methods in Urban Development: Some Examples and Implications," Geological Society of America Spec. Paper No. 174, pp. 1–35 (1976).

Hauser, J. R. "Use of Water Hyacinthoquatic Treatment Systems for Ammonia Control and Effluent Polishing," *Journal of Water Pollution Control Federation*, 56(3):219–225 (1984).

Hipel, K. W. et al. "Political Resolution of Environmental Conflicts," *Water Resources Bull.*, 12(4):813–828 (1976).

Howells, D. H. "Land Use Functions in Water Quality Management," *Water Resources Bull.*, 7(1):162–170 (1971).

Jackson, B. B. "The Use of Streamflow Models in Planning," *Water Resources Res.*, 11(1):54–63 (1975).

James, L. D. "Formulation of Nonstructural Flood Control Programs," *Water Resources Bull.*, 11(4):688–705 (1975).

Koshal, R. K. "Urban Growth and Environment: Reply," *Growth and Change*, 7(4) (1976).

Levin, M. R. "Why Can't Johnny Plan?" *Planning*, 42(8):21–23 (1976).

Logan, J. A. et al. *Environmental Engineering and Metropolitan Planning*. Northwestern University Press, 265 pp. (1962).

McGrath, D. C., Jr. "Planning: Some Questions, Answers and Issues," *J. Soil Water Cons.*, Jan./Feb., pp. 6–38 (1973).

McLoughlin, J. B. *Urban and Regional Planning: A Systems Approach*. London:Faber and Faber, 331 pp. (1969).

Mayer, H. M. "The Future of Cities: Governmental Coordination and Planning," *J. Geog.*, 68(9):518–526 (1969).

Mudroch, A. and J. A. Capobianco. "Effects of Treated Effluent on a Natural Marsh," *Journal of the Water Pollution Control Federation*, 51(9):2243–2256 (1979).

Peterson, W. "On Some Meanings of 'Planning,'" *J. Am. Inst. Planners*, 32(3):130–142 (1966).

Reilly, W. K. *The Use of the Land: A Citizen's Policy Guide to Urban Growth*. New York:Thomas Y. Crowell, Co., 318 pp. (1973).

Reimer, P. O. and J. B. Franzini. "Urbanization's Drainage Consequence," *J. Urban Plann. & Dev. Div.*, ASCE, 97(UP2):217–237 (1971).

Rose, J. G. "Legal Foundations of Land Use Planning: Cases and Materials on Planning Law," Urban Policy Research, Rutgers University, New Brunswick, NJ (1974).

Schultz, N. U. and A. Wilmarth. "Water Quality Simulation and Public Law 92-500. Case Study: Southwestern Illinois," *Water Resources Bull.*, 14(2):275–287 (1978).

Stowell, R. et al. "Concepts in Aquatic Treatment System Design," *Journal of the Environmental Engineering Division*, American Society of Civil Engineers, 107(EE5):919–940 (1981).

7 | Structural Control Measures

THIS CHAPTER WILL outline a few of the major structural alleviative measures. The employment of each one of these measures must be judged according to its merits in meeting the climatic and physiographic needs of a region. Accordingly, each measure has its advantages and disadvantages, and not all are suitable for every region. Most likely, the optimum approach would be to utilize several methods in various degrees. The point of this chapter is that the technology is available to substantially reduce water quality and water quantity problems which now exist in urban areas.

7.1 Water Quantity

Figure 2.3 demonstrated that the increased imperviousness associated with urban areas facilitated rapid runoff of storm waters. In essence, the storage of rainwater is decreased in this manner. Alternatively, if one wishes to reverse this process, i.e., to decrease the rate and/or volume of runoff, one may increase the storage factor. This may be accomplished by the methods outlined below.

ROOFTOP STORAGE

Chiang (1971) pointed out that roofs make up a large proportion of the impervious area within an urban region. He concluded that more than 50% of the impervious area in a city was occupied by roofs, and in a typical middle-class residential area, roofs could make up 10–25% of the impervious area. If one were to design roofs to retain rainwater (up to a foot deep), a considerable amount of water could be held in storage and slowly released after the storm passed.

The author points out that roofs are not subjected to littering as urban streets are, and therefore, the quality of rainwater is very good, depending of course on air pollutants. In many countries (e.g., Taiwan), roof water is "harvested" purposefully, for cooking, drinking and washing. In Chiang's words:

> . . . instead of using [a] roof as a drainage device as we are doing now, why can't we employ it as a control and regulating device, as well as a conservation measure? (p. 172).

Design requirements would not be cost-prohibitive. Chiang states that presently, roofs are commonly built to hold eight inches of water (40 lb/ft^2), and an additional four inches would not add significantly to construction costs, because the cost of a roof is a small fraction of the total cost of the majority of urban structures. Roofs should be designed flat with a gentle slope for drainage. The downspout should be capable of conveying six inches of rainwater in at least two days, and may be connected into a storm sewer line or into a previous groundwater recharge pit. Roof water may be harvested and may augment existing city supplies, either directly or by subsurface flow (via recharge pits).

Perhaps the most significant economic factor which Chiang stresses is that the detention of rainwater by roofs could considerably reduce the size of a city's storm sewerage system. He suggests that since rooftop detention could detain up to 50% of the peak discharge, it could be a determining factor in storm sewerage design. In other words, sewerage which was designed to convey a 20-year storm could, with the aid of rooftop detention, safely handle up to a 100-year event. In this manner, a city could initially install smaller sewers and save substantially in construction and in long-term maintenance costs.

DeCook and Foster (1984) suggest the harvesting of rainwater in the Tucson, Arizona urban area. They point out that besides augmenting residential and municipal water supplies, and recharging groundwater, there would also be a reduction in flood hazards or peaks.

POROUS PAVEMENT

Thelen et al. (1972) investigated the potentials of various porous pavements to increase perviousness, allow infiltration of storm waters, and thereby reduce surface runoff. The authors found, for various economic and physical reasons, that an open-graded[58] asphalt concrete was the most suitable material. This material exhibited superior physical characteristics, was low in cost, and could be laid by conventional paving methods.

Three types of asphaltic concretes were analyzed. Infiltration rates for these

---

[58]"Open-graded—aggregate containing relatively small amounts of fine particles. Generally contains one 10% by volume of voids in an asphalt concrete" (Thelen et al., 1972, p. 142).

concretes varied from 5 in./hr to 25 in./hr. Thelen et al. define this porous asphaltic concrete as:

> ...a graded aggregate cemented together by asphalt cement into a coherent mass, with sufficient interconnected voids to show a high permeability to liquid water. . . . In manufacture, the graded aggregate is *derived* by heating to 275–300°F, and mixed with hot asphalt cement. The hot-mix material is hauled to the paving site, where it is applied in a spreading machine in smooth layers, and compacted to design density by heavy rollers (p. 25).

The authors outline several benefits attributed to the open-graded asphaltic surface. Construction of the porous pavement over an impervious surface would allow detention-storage of storm waters and, where possible, the waters could be released at a later date. Storage could augment existing water supplies, or, if the porous asphalt were placed over a permeable land surface, the waters could recharge urban aquifers, which often are heavily drawn upon. Retention of storm waters would alleviate flash flooding and preserve natural drainage patterns. Conservation of storm runoff waters is becoming an increasingly important issue, especially in cities in drier climates (Angino et al., 1972). The temporary detention of storm waters would also have a beneficial effect on improving surface runoff water quality by reducing the number of shock loadings.

Thelen et al. point out that research completed on airport runways equipped with similar asphaltic surfaces shows that these surfaces have successfully prevented accidents due to skidding and hydroplaning. Also, this type of pavement enhances the visibility of pavement markings, which, the authors claim, adds to overall safety.

Costs of installation and maintenance of the porous asphaltic surface is equal to or cheaper than the cost of conventional pavement with storm or combined sewer facilities.

SURFACE DETENTION

Field, Tafuri and Masters (1977) report that surface ponding is the most common form of detention employed by developers. Table 7.1 shows the differences in costs between this method and conventional sewerage.

Developers may easily adapt a parking lot to a surface pond by proper grading in conjunction with curbing installed to retain the runoff of waters. If porous pavement is utilized in specific areas of the parking lot, the impounded runoff may not even pose a problem to pedestrians or vehicles. In the past, the concept has always been to grade parking lots and design sewerage which together would allow storm waters to run off rapidly. As a consequence, these waters have caused much concern, as has been discussed in Chapters 2 and 3. If this concept is modified somewhat, a detention reservoir may be created which will attenuate rapid runoff.

Table 7.1. Cost comparison between surface ponding techniques and conventional sewer installation (R-8).

		Cost Estimate ($)	
Site	Description	With Surface Ponding	With Conventional Sewers
Earth City, Missouri	A planned commmunity including permanent recreational lakes with additional capacity for storm flow	2,000,000	5,000,000
Consolidated Freightways, St. Louis, Missouri	A trucking terminal using its parking lots to detain storm flows	115,000	150,000
Ft. Campbell, Kentucky	A military installation using ponds to decrease the required drainage pipe sizes	2,000,000	3,370,000
Indian Lakes Estates, Bloomington, Illinois	A residential development using ponds and an existing small diameter drain	200,000	600,000

Source: Field et al. (1977), p. 31

STORM WATER DETENTION STRUCTURES

Nassau County on Long Island, New York, developed a long-range drainage plan in 1935 in which the county sanitation commission opted to construct recharge basins instead of storm sewer trunks (Seaburn, 1970). This decision was implemented partly because it was believed that it was less expensive to construct and maintain recharge basins than storm sewerage.

A system of short sewer lines, terminating in open basins or pits, was constructed throughout the county. About 500 recharge basins are maintained in Nassau County (adjacent Suffolk County has about 1400). The average size of a basin is two acres and the range is from 0.1 to 30 acres, with an average depth of 10 feet. They are unlined pits, and are excavated below the land surface in mostly moderate to high-permeable sand and gravel deposits of glacial outwash. Most recharge basins are located well above the water table.[59]

The basic design criteria for a basin has evolved on a trial and error basis since 1935. There are two general types of basins: those with overflow structures and those without. An overflow structure will convey to a nearby stream or another basin (by pipes, flumes or gutters) runoff that exceeds the capacity of the basin. Basins with overflow structures have two main design criteria.

---

[59]See Singh (1976) for a discussion of subsurface flow under a recharge basin.

The first is the volume of the basin below the overflow elevation which is estimated by multiplying (a) the volume of water derived from a 5-in. rainfall over the total drainage area of the watershed by (b) a factor ranging from 40 to 90%. The factor depends on land slope and percentage of area occupied by streets and parking lots (40% is used in most residential areas, and industrial areas may attain 90%). The second design criterion concerns the elevation of the overflow structure, which must not be more than 10 feet above the floor of the recharge basin. In the design, infiltration into the floor and sides of the basin is neglected to give an extra factor of safety.

Overton and Meadows (1976) suggest that a storm water detention basin could amount to little more than the water backup behind a highway or road culvert. In their example, a 36-inch concrete culvert goes under a roadway. When inflow exceeds the capacity of the culvert, water becomes impounded as much as 2.17 acre-feet (3.15 feet of headwater) before inflow will overtop the road. (The authors describe the design of a basin for an 85-acre residential watershed in Knoxville, Tennessee.) It is obvious that many such intentional or nonintentional detention structures exist such as this one.

Abt and Grigg (1978) propose methods of design for detention reservoirs. They caution that detention reservoirs should not be placed haphazardly in a catchment, because their effects could aggravate rather than alleviate potential flood hazards. In their design, they use a triangular hydrograph (Figure 7.1) to compute volume, along with the following formula:

$$\text{Volume} = \frac{1}{2} Q_p (T_c + mT_c) \frac{60}{43560}$$

where

Volume = the approximate volume of storm runoff
Q_p = peak discharge in cfs

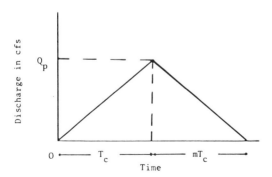

Figure 7.1. Triangular hydrograph [*Source*: Abt and Grigg (1978), *Water Resources Bull.*, 14(4): 956–965. The American Water Resources Assoc., Minneapolis, MN].

218   STRUCTURAL CONTROL MEASURES

T_c = time of concentration or lagtime
mT_c = the recession duration

When there is a series of detention reservoirs within a watershed, the volume computations become more difficult, as the authors point out.

Sewers may be designed to store storm waters. The city of Chicago has constructed storm water storage tunnels under the city.

NEW DESIGN OF CITY BLOCKS

If urban areas were designed to preserve the natural environment, rather than being designed to totally replace it, then many of the runoff problems associated with impervious areas could be eliminated. In discussing the work of others, and by presenting his own ideas, Martin (1972) suggests the latter. He points out the equal areal principle demonstrated by Fresnel's diagram. If one were to compute the area of each successive annular ring in Figure 7.2, one would find that even though each ring diminishes in width, it has exactly the same areas as its predecessor.

The proper application of this principle in building design and layout on the land surface could greatly preserve the natural environment. Figure 7.3 demonstrates such an application; the amount of floor space remains the same in either layout. The right layout shows that by far the majority of the land use is made up of natural surface.

7.2 Water Quality

Storage is also of central focus in water quality problems generated by urban area, however, in the opposite manner from water quantity. The accumulation and storage of urban residue and dust on surfaces has been shown in Chapter

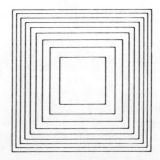

Figure 7.2. Equal area principle demonstrated by Fresnel diagram [*Source*: Martin, L. and L. March, eds. (1972), *Urban Space and Structures* (New York:Cambridge University Press), p. 19].

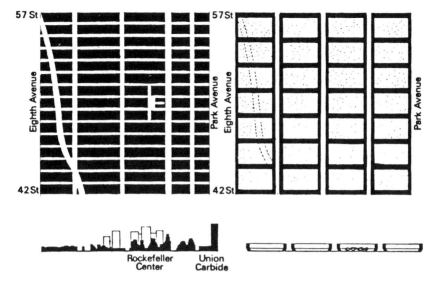

Figure 7.3. Application of equal area principle [*Source*: Martin, L. and L. March, eds. (1972), *Urban Space and Structures* (New York:Cambridge University Press), p. 21].

3 to have a shock-loading effect on streams. There are two basic structural approaches to reduction of shock loads: (1) incremental removal of the accumulated materials, thereby spreading the impact of shock loads over a much wider period of time, and (2) treatment of storm water runoff by sewage treatment plants.

INCREMENTAL REMOVAL

In Section 5.6, the effects of street sweeping were briefly discussed. After a model study of an area in Washington, DC, Sutherland and McCuen (1978) suggested that vacuumized sweepers seemed to be 5% more efficient than motorized ones, and that sweepers operating at 3 mph were about 3% more efficient than ones operating at 4 mph. The most efficient street sweeping operation, for the four areas studied, was found to be by a 3-mph vacuumized sweeper at an interval of once every two days. These observations could not be considered to hold true for all regions, but, nevertheless, frequency of sweeping and speed and type of street-sweeping equipment have a direct bearing on the efficiency of the cleaning effort.

Sartor and Boyd (1977) acknowledged that:

> . . . acceptable methods of planning and evaluating the efficiency of street cleaning programs are not available at the present time (p. 47).

They recommended that:

> . . . public works departments maintain accurate and detailed records of street cleaning operations, including manpower utilization, equipment utilization, and equipment maintenance (p. 48).

In this manner, an analysis of the data would provide a means to arrive at a cost-effective street cleaning program.

Field and Lager (1975) reported that the efficiency of street sweepers to remove the dust and dirt fraction (contributing most heavily to water pollution loadings) average 50%, whereas the pickup of litter and debris ranged from 95 to 100%. They also recommended the use of the more efficient vacuums which pick up the fine particles. They listed a cost range of $2.18/curb mile to $8.42/curb mile swept.

Sartor and Boyd observed that pavement type and condition affected the availability of loose particulate matter found on streets. They found that all-concrete streets were cleaner than all-asphalt streets, and mixed concrete-asphalt streets were intermediate. In any of the three cases, streets maintained in good condition were substantially cleaner than those in fair to poor condition. Sartor and Boyd recommended that public works departments pay increased attention to maintaining pavements in good condition.

TREATMENT OF STORM WATER DISCHARGE

There are two obvious problems with the concept of treatment of storm water discharge. The first is that all urban runoff does not enter the sewerage system, and streams will invariably receive some untreated runoff. The second problem concerns the capacity of the sewage treatment plant. When inputs exceed the plant's processing capability, some plants will automatically bypass the inflow into a stream. Other factors which complicate sewage treatment plant designs are outlined by Field and Lager (1975):

> . . . also, due to the intermittency and variability of stormwater runoff and interrelated system flows, there is no such thing as an "average" design condition for stormwater treatment facilities . . . in addition, the magnitude, debris content and brute force of storm flows may be major constraints limiting options for centralization (because of high transmission costs) and rendering sophisticated and complex equipment ineffectual or impossible to maintain (p. 109).

An alternative to the above may be provided by the collection and storage of the storm water. In this manner, in a later nonstorm period, the sewage treatment plant operator can feed the impounded water through the treatment facilities at a controlled rate of flow. Pitt and Field (1977) point out one disadvantage of this method; the costs of storage facilities in heavily urbanized areas may be $25,000 per hectare, or $10,000 per acre served.

Field and Lager point out that physical treatment processes are in many ways well-suited to storm water applications. These processes consist of the conventional sedimentation, dissolved air flotation, screening, filtration and special regulator devices.[60] Sensitivities of flows and loads may be a problem in some cases, as well as the effective utilization of the facilities between storm periods.

Conventional secondary sewage treatment plants usually consist of: (1) mechanical filtering, (2) sedimentation and aeration facilities, and (3) biological filtering. For some time now, scientists have studied the possibilities of expanding the biological filtering concept and its potential to treat raw sewage. In sewage treatment plants, the biological filter consists basically of a bacterial biomass, whereas some scientists have slightly modifed this concept and have studied the use of aquatic plants, or a basically botanical filtering system (see Seidel, 1976).

Wolverton et al. (1976) wrote:

> ... vascular aquatic plants, when utilized in a controlled biological system including a regular program of harvesting to achieve maximum growth and pollution removal efficiency, may represent a remarkably efficient and inexpensive filtration and disposal system for toxic materials and sewage released into waters near urban and industrial areas (p. 141).

The authors report that under favorable conditions, one acre (0.40 ha) of water hyacinths can: (1) produce over 534 lb (240 kg) of dry plant material per day, (2) remove 3500 lb (1591 kg) of nitrogen and over 800 lb (364 kg) of phosphorous annually from sewage effluent, and (3) absorb over 150 lb (68 kg) of phenol every 72 hr, and 120 g of trace heavy metal contaminants every 24 hr.

Seidel (1976) has studied the use of *Schoenoplectus lacustris palla* (bulrushes) in the purification of water. This plant species adapts well to polluted waters. The plants display an amazing capability not only to eliminate dangerous industrial poisons (phenol) but also to gain in biomass. The plants' capabilities were tested with other poisons such as chlorophenol pentachlorophenol and cyanogen. In some cases, a cascade system was established, and the author reports very satisfactory results. These poisons pose a serious threat to conventional sewage treatment plants, in that they kill the biological filters. If discharged into a stream, they greatly upset the aquatic ecology and in general have always posed a disposal problem.

Seidel goes on to propose the use of the sun and gravity as a new type of sewage treatment plant. In the author's words:

> ... today, we make the most complicated efforts to capture and use the energy of the sun. The green plant is an age-old model that we cannot respect and value too highly ... gravity can be used in the transport and aeration of liquid wastes

---

[60]See Culp (1976) for excellent discussions and illustrations on wastewater treatment.

through the design of sloping trenches arranged in a sequence on varying levels (p. 119).

Richardson and Daigger (1984) present the results of an investigation of an active solar-aquaculture demonstration facility in Hercules, California. This plant provided secondary treatment to raw municipal wastewater. The system used water hyacinths placed in an "aquacell." Failure of the facility was considered to be the result of process and design deficiencies. Richardson and Daigger point out:

> As with all technologies, aquaculture should be applied only in the proper situation. In the writers' opinion, it is best used as a secondary treatment or polishing step for wastewaters that have received at least some degree of prior treatment (p. 960).

In the last few paragraphs, the results of some research in the use of biological means to treat raw sewage have been presented. These methods have been used and tested primarily on point sources. Nonpoint sources are yet another matter, requiring further research and design. However, the technology is presently available and the principles remain essentially the same. The major obstacles to be overcome are flow- and climate-related. Stormflows are variable and their stochastic nature complicates the design and maintenance of botanical filters for the periods between storms. Climate has a direct bearing on the growth and survival of aquatic plants, and any botanical filter must be designed with this in mind. It seems the most rational approach to the solution of a nonpoint source pollution treatment problem will ultimately be an integrated one. Most likely, the use of a physical treatment system along with the use of storage of storm waters, and treatment by a botanical filter will achieve a sound design scheme.

The accumulation and storage of surface runoff in sewers has been shown to cause a shock-loading effect on streams after being flushed out by storm water. Because this is a storage problem its means of solution is the same in principle as those proposed at the beginning of this section, i.e., incremental release and/or treatment. Incremental release of these concentrated constituents may be accomplished by the construction of storage facilities and consequent controlled release (see Figure 7.4). Water treatment may be a part of release design or treatment may consist of in-sewer mechanical devices designed to aerate the accumulated materials and reduce their concentration. Field et al. (1977) suggest different sewer designs which would discourage the sedimentation of solids and provide flow for solids at lower velocities.

The focus of Public Law 92-500 has been to minimize or, hopefully, to eliminate water pollutant discharges to streams. Streams do have the capability of assimilating a certain amount of waste effluents. If, in some mechanical way, the assimilative capability of a stream could be enhanced, then pollutant dis-

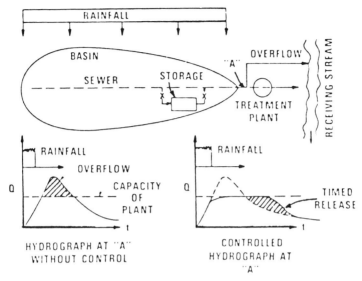

Figure 7.4. Results of controlling stormflow by storage [*Source*: Field et al. (1977), p. 46].

charges could be increased or at least allowed. Sewers do not collect all urban runoff, and pollutants do find their way to rivers. Hopefully, these nonpoint source pollutants are small in volume and concentration. In any case, enhancement of the receiving stream's assimilative capacity could, in principle, aid in the overall water quality planning for an urban area.

One such approach would be to allow nonpoint and point source discharges with minimal treatment into urban streams. In this case, we will consider primarily organic discharges with some inorganics, but no industrial toxics. These streams would become controlled septic regions; within these regions, no contact recreation would be allowed. Downstream from these zones would be mechanical aerators placed under the stream, capable of totally stabilizing the inflow and allowing the downstream area to be available for contact recreation. In order to implement this approach, the scientist would have to have an intimate understanding of the river of concern, the local climate and the potential growth and decay rate of the bacteria involved. However, the technology does exist and, in principle, a water resource method similar to this is being implemented in Germany.

7.3 Sheet Erosion and Sedimentation Control

Sheet erosion may easily occur at sites where vegetation is removed and the soil surface lies exposed to the elements as well as further mechanical agitation by excavation vehicles. The processes of sheet erosion are shown in Figure 7.5. Raindrops which achieve velocities of 19 mph (Environmental Protection Agency, 1971) strike the soil surface, impart a considerable amount of kinetic energy on the surface and separate soil particles. As more rain follows, the surface becomes saturated and overland flow begins, carrying the water–soil particle mixture. During the transportation process, the mixture in turn may abrade and accumulate more soil particles as it travels and adds to the overall concentration of the water–soil particle mixture. As the various flows accumulate, downcutting is initiated, and rill and gully erosion occurs. Sedimentation begins when soil particles settle out of the mixture. This is a result of decreases in transport energies, and may be caused by smoother slopes and/or ponding.

Thronson (1973) gave a brief history of the development of the Erosion Index (EI). Musgrave (1947) attempted to quantify the loss of soil by erosion by considering slope and length of agricultural lands. In 1958, the universal soil loss equation (a refinement of Musgrave's work) was developed by Wischmeier, Smith and Uhland, and is currently used by the U.S. Soil Conservation Service, and is as follows (Swerdon and Kountz, 1973):

$$A = RKLSCP$$

where

A = computed soil loss in tons per acre
R = rainfall factor
K = soil erodibility factor
L = slope length factor
S = slope gradient factor
C = cropping-management factor
P = erosion control practice factor

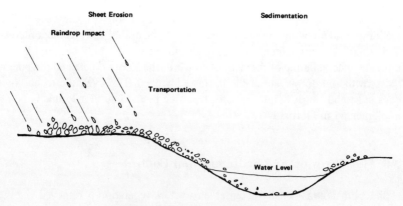

Figure 7.5. The processes of sheet erosion.

Evaluative methods for these various factors as well as their proper employment may be found in U.S.D.A. Agricultural Research Service Agriculture Handbook No. 282, or later and revised versions (see Wischmeier and Smith, 1965). Swerdon and Kountz (1973) discuss the determination of these various factors within the state of Pennsylvania, and may provide some insight into the use of this equation in other areas.

PREVENTION OF SHEET EROSION

Observing Figure 7.5, if one strives to eliminate raindrop impact, then sheet erosion may be greatly limited. This may be accomplished by a minimal removal of the vegetative canopy and/or by placing various artificial covers in the path of the raindrops so that their energies are dissipated before they arrive at the soil surface.

Thronson (1973) lists several methods of providing surface cover. Excelsior blankets are machine-produced mats of curled wood excelsior consisting of 80% 8-in. or longer fiber length. These mats, along with other types of mats, may be stapled to the ground surface. Fiber mulches, hay, straw or wood chips may be employed quite effectively; however, if sheet erosion should start, these dissipators soon lose their effectiveness. There are chemical soil stabilizers which serve to maintain surface integrity and prevent sheet erosion. Quick-growing grasses have been used in various climates, and offer a new protective "natural" canopy. These grasses may be applied along with fertilizer in one blowing operation.

Transportation of the water–sediment mixture may be limited by methods of separating the soil particles from the water and/or dissipation of transport energies. This is accomplished by filtering using straw bales, sand and gravel. The water flows through these filters, meeting with some resistance, which thereby dissipates energy; at the same time the larger sediment particles are held back and are deposited before the filters.

Prevention of rill and gully erosion may be accomplished by several methods. The first is by use of filters of small semipervious barriers (EPA, 1971) which serve as energy dissipators and particle separators. Sheet flow may be purposefully concentrated; to eliminate downcutting by this intensified flow a rip-rapped and/or concrete-lined channel, and/or a flexible slope drain (a very large hose) and an energy dissipator of sand and gravel are employed to remove the sheet flow (see Figure 7.6). The concept behind all these latter measures is to dissipate the downcutting energies of the intensified flow.

Sedimentation may be largely eliminated by the proper employment of the preceding methods.[61] On the other hand, sedimentation may be allowed to oc-

---

[61]For a good overall review of the effectiveness of several sediment control techniques, see Reed (1978).

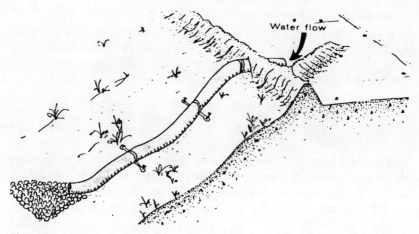

Figure 7.6. Temporary, flexible slope drain. Discharges on gravel energy dissipator to prevent erosion at discharge end [*Source*: EPA (1971), p. 36].

cur on the construction site with the objective that after the construction activities are completed, the sediment will remain at the site and not affect downstream natural systems. Locally enhanced sedimentation may be accomplished through the use of sediment retention basins (see Figures 7.7 and 7.8). These are usually located at the lowest point of the construction site and consist of a manually excavated pond preferably with a drain as shown in Figure 7.7. The objective here is to provide standing water which will allow the soil particles to settle, and not allow the basin to fill and spill over, carrying the sediment–water mixture to nearby streams. Sediment retention basins may be used in conjunction with rill and gully erosion prevention methods. At the end of the construction project, the sediment may be dredged from these ponds, or in some cases the pond is allowed to remain and may enhance the natural aesthetics of the new development, as well as to continue to function as a sediment retention structure.

Costs of the various sheet erosion and sedimentation preventative measures are discussed in detail by Thronson (1973). Today, his cost figures would be higher because inflation continues to add on to these estimates. Nevertheless, a sound insight into the costs of sheet erosion control may be obtained from his estimates. For comparative purposes, Figure 7.8 gives an illustration of a sediment and erosion control plan for a construction site and Table 7.2 lists the suggested 1973 costs.

The EPA (1971) estimates that along a highway with an average construction cost of $1,000,000 per mile, it would cost between $10,000 and $15,000 per mile for erosion and control measures. Housing developments average about $40 per lot (estimated by engineering and geologic consultants) and $100 per lot (as estimated by developers).

7.4 Summary

WATER QUANTITY

Chapters 2 and 3 outline the impact of urban areas on storm water quantity and storm water quality. The component of the hydrologic cycle that is reduced by urban areas is storage, which influences quantity, which interacts with quality, etc. Hence, in concept, the solution to the storage problem is simple; reverse the rapid runoff process by increasing storage.

Chiang (1971) suggests, because roofs make up such a large part of the impervious portion of urban areas, storing rainwater on roofs and releasing it after a storm. This is perhaps the most sound of the water quantity approaches presented; when the water is impounded on roofs, there most likely would not be any water quality impairment. Water quality may degrade in surface and/or subsurface detention reservoirs. Thus, the usual water quantity and water quality problems associated with urban runoff would be eliminated.

Thelen et al. (1972) studied the potentials of porous pavements, i.e., making the ground less impervious in an attempt to increase storage. This would appear to be a feasible method because costs of installation would be no more than those of conventional pavements.

Substantial savings may be achieved by developers in the use of surface ponding over the installation of conventional sewers as Field, Tafuri and Masters (1977) point out. If porous pavements are specifically situated in certain areas of the pond, inconveniences (i.e., stepping in puddles, driving though puddles, etc.) to pedestrians and automobilists may be greatly reduced.

For some time, recharge basins have been used to impound storm waters in Nassau County, New York. These basins may have overflows which conduct high flows to other basins or to a nearby stream. In some cases, water quality in detention basins may deteriorate and, although a water quantity problem is solved, a serious water quality problem is developed. In order to reduce this problem, the basins require periodic maintenance. Basins are, in the long run, usually cheaper to maintain than regular sewers and may serve to recharge heavily depleted urban aquifers.

Figure 7.7. Small sediment basin with outlet pipe discharging on energy dissipator to prevent erosion at discharge end [*Source*: EPA (1971), p. 40].

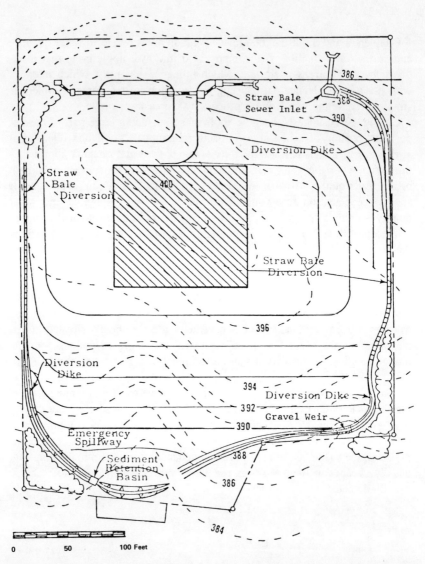

Figure 7.8. Example of sediment and erosion control plan [*Source*: Thronson (1973), p. 120].

Table 7.2. Estimating cost of sediment and erosion control plan.

Item	Unit	Quantity	Unit Cost ($)	Cost California ($)	Cost Virginia ($)
Straw Bale Sewer Inlet	each	1	55.00 (46.34)	55	46
Diversion Dike	L. Ft.	575	4.51 (3.70)	2593	2127
Straw Bale Diversion	Bale	90	7.86 (6.62)	707	596
Gravel Weir	6 ft	2	10.44 (8.99)	221	18
Sediment Ret. Basin	cu yd	36	16.25 (13.75)	585	495
Total Cost for 2.95 acre project				3961	3232
Cost per Acre[a]				1340	1110
If hydromulch & seeding is used on sloping banks, add for:					
California				720	
Virginia					680
Grand Total Cost Per Acre				2060	1790

[a]Rounded off to nearest ten dollars.
Source: Thronson (1973), p. 121.

Overton and Meadows (1976) point out the detention that occurs behind a highway culvert. Abt and Grigg (1978) show methods of design of detention reservoirs and caution that they should not be haphazardly placed in a catchment, for detention reservoirs may aggravate rather than alleviate flooding.

If city blocks are designed and constructed to preserve the natural environment, rather than totally replace it, then urban runoff problems could be greatly reduced. Martin (1972) develops a few concepts dealing with the use of successive annular rings, each containing equal areas, and all within the same rectangle. The outcome is that by constructing buildings optimizing on the annular ring concept, a modified city block is created, the same floor area is maintained, and the majority of the natural landscape remains intact. In this manner, the natural hydrology of the area is only slightly modified.

The accumulation and storage of urban refuse and dust on surfaces has been shown in Chapter 3 to create a shock-loading effect on streams. There are two approaches to solving this problem: incremental removal of the accumulated debris and treatment of storm water runoff.

Present street-sweeping operations are aimed to make the urban environment more aesthetically pleasing, and do not substantially reduce pollutant loads to streams. This fact is noted by Field and Lager (1975) when they note the low efficiencies of street sweepers in removing the dust and dirt fraction. Sutherland and McCuen (1978), after developing and employing a model on hypothetical locations in the Washington, DC area, suggested that the most efficient street-sweeping operation was to use a 3-mph vacuumized sweeper at an interval of once every two days.

Sartor and Boyd (1977) request that public works departments maintain accurate and detailed records of street-cleaning operations in order that a data base may be established from which research may be conducted in order to define cost-effective approaches.

Adequate treatment of storm water discharges suffers from a few problems. First, all urban runoff does not enter the sewerage system and runoff pollution to urban streams will inevitably occur. Second, inflows from storms may exceed the sewage treatment plant's capabilities. Field and Lager (1975) point out that average design conditions for such inflows probably do not exist. An alternative is to collect and store inflows. However, the costs of constructing impoundment facilities may be prohibitive.

Seidel (1976) has demonstrated the capabilities of bulrushes (*Shoenoplectus lacustris palla*) to purify water. These plants display an amazing ability to process poisons and yet to gain in biomass. Seidel proposes the design and construction of a more natural sewage treatment plant employing the sun and gravity. A major limitation to botanical filters is that they depend on climate. However, they may be used together with other filters in an integrated sewage processing network.

If a river were partitioned into zones, each one maintained according to certain water quality standards, urban nonpoint and point source discharges could be allowed. This would create the development and maintenance of controlled septic regions. Downstream of these regions would be mechanical aerators capable of stabilizing the septic flow and, for example, permitting an area downstream with water clean enough for contact recreation. Such a scheme may be a practical and economical alternative to other methods now employed by urban areas.

EROSION CONTROL

At construction sites where vegetation is removed and the soil surface lies exposed to the elements, erosion may easily occur. The universal soil loss equation may be applied to estimate the amount of erosion.

Erosion prevention may be accomplished by providing a canopy over the soil. Fiber mulches, hay, straw or wood chips may be used to dissipate the energy of incoming rainfall (Thronson, 1973). If the water–soil particle mixture should begin to move, filters such as straw bales may be employed to separate the water from the sand and gravel. The bales also act to dissipate energy from the waters and thereby diminish the erosive capabilities of the water downslope.

Sedimentation basins impound the storm waters from a site and allow the sediment to settle instead of flushing into nearby streams.

The costs of these various measures are negligible in comparison to the environmental damage that might occur if they are not implemented.

7.5 Exercises

Considering an urban area with which you are very familiar:

1 What structural control measure should be undertaken to alleviate what you feel to be the major urban hydrologic problem related to water quantity? Defend your choice.
   a) Could the problem have been prevented if one or more of the storm water alleviative methods outlined in Section 7.1 had been incorporated in the original design of the municipality? How so?
2 What structural control measure should be undertaken to alleviate what you feel to be the major urban hydrologic problem related to water quality? Defend your choice.

References

ABT, S. R. and N. S. Grigg. "An Approximate Method for Sizing Detention Reservoirs," *Water Resources Bull.*, 14(4):956–965 (1978).

ANGINO, E. et al. "Effects of Urbanization on Storm Water Runoff Quality: A Limited Experiment, Naismith Ditch, Lawrence, Kansas," *Water Resources Res.*, 8:135–140 (1972).

CHIANG, S. L. "A Crazy Idea on Urban Water Management," *Water Resources Bull.*, 7(1):171–174 (1971).

CULP, G. "Environmental Pollution Control Alternatives: Municipal Wastewater," U.S. EPA Technology Transfer No. EPA-625-76-012, 79 pp. (1976).

DECOOK, J. K. and K. E. Foster. "Systems for Rainfall and Runoff Use, Tucson, Arizona," *Water Resources Bulletin*, 20(6):883–887 (December 1984).

ENVIRONMENTAL PROTECTION AGENCY, OFFICE OF WATER PROGRAMS. "Control of Erosion and Sediment Deposition from Construction of Highways and Land Development," Division of Technical Support, Technical Assistance Branch, Rural Wastes Section, 50 pp. (1971).

FIELD, R. and J. A. Lager. "Urban Runoff Pollution Control—State-of-the-Art," *J. Env. Eng. Div.*, ASCE (EE1):107–125 (1975).

FIELD, R. et al. "Urban Runoff Pollution Control Technology Overview," EPA Technology Series EPA-600-2-77-047, 91 pp. (1977).

MARTIN, L. "The Grid as Generator," in *Urban Space and Structures*. L. Martin and L. March, eds. New York:Cambridge University Press, Chapter 1 (1972).

MUSGRAVE, G. W. "The Quantitative Evaluation of Factors in Water Erosion: A First Approximation," *J. Soil Water Cons.*, 2:133–138 (1947).

OVERTON, D. E. and M. E. Meadows. *Stormwater Modeling.* New York: Academic Press, Inc., 358 pp. (1976).

PITT, R. and R. Field. "Water Quality Effects from Urban Runoff," *J. Am. Water Works Assoc.*, 69(8):432–436 (1977).

REED, L. A. "Effectiveness of Sediment-Control Techniques Used During Highway Construction in Central Pennsylvania," U.S. Geological Survey Water Supply Paper 2054, 57 pp. (1978).

RICHARDSON, D. L. and G. T. Daigger. "Aquaculture: The Hercules Experience," *Journal of Environmental Engineering Division*, American Society of Civil Engineers, 110(5):949–960 (1984).

SARTOR, J. D. and G. B. Boyd. "Water Pollutants in Urban Runoff," *J. Am. Water Works Assoc.*, 69(8):432–436 (1977).

SEIDEL, K. "Macrophytes and Water Purification," in *Biological Control of Water Pollution.* J. Tourbier et al., eds. Philadelphia: University of Pennsylvania Press, Chapter 14 (1976).

SEABURN, G. E. "Preliminary Results of Hydrologic Studies at Two Recharge Basins on Long Island, New York," U.S. Geological Survey Professional Paper No. 627-C, 17 pp. (1969).

SINGH, R. "Prediction of Mound Geometry Under Recharge Basins," *Water Resources Res.*, 12 (4):775–780 (1976).

SUTHERLAND, R. C. and R. H. McCuen. "Simulation of Urban Nonpoint Source Pollution," *Water Resources Bull.*, 14(2):409–428 (1978).

SWERDON, P. M. and R. Kountz. "Sediment Runoff Control at Highway Construction Sites," Engineering Research Bulletin No. B-108, The Pennsylvania State University, College of Engineering, University Park, PA, 70 pp. (1973).

THELEN, E. et al. "Investigation of Porous Pavements for Urban Runoff Control," prepared for the Office of Research and Monitoring, Environmental Protection Agency, Contract No. 14-12-924, 142 pp. (1972).

THRONSON, R. E. "Comparative Costs of Erosion and Sediment Control, Construction Activities," U.S. EPA, Washington, DC, 205 pp. (1973).

THRONSON, R. E. "Nonpoint Source Control Guidance Construction Activities," U.S. EPA, Office of Water Planning and Standards, Washington, DC (1973).

WISCHMEIER, W. H. and D. D. Smith. "Predicting Rainfall-Erosion Losses from Cropland East of the Rocky Mountains, Guide for Selection of Practices for Soil and Water Conservation," U.S. Department of Agriculture, Agricultural Research Service, in cooperation with Purdue Agricultural Experiment Station, Agriculture Handbook No. 282, 47 pp. (1965).

WISCHMEIER, W. H. et al. "Evaluation of Factors in the Soil-Loss Equation," *Agric. Eng.*, 39: 458–462 (1958).

WOLVERTON, B. C. et al. "Application of Vascular Aquatic Plants for Pollution Removal, Energy and Food Production in a Biological System," in *Biological Control of Water Pollution.* J. Tourbier, et al., eds. Philadelphia: University of Pennsylvania Press, Chapter 17 (1976).

Selected Readings

Carpenter, R. L. et al. "Aquaculture as an Alternative Wastewater Treatment System," in *Biological Control of Water Pollution.* J. Tourbier et al., eds. Philadelphia: University of Pennsylvania Press, Chapter 24 (1976).

Evenson, N. "LeCorbusier: The Machine and the Grand Design," New York: George Braziller, 128 pp. (1969).

Selected Readings

Finnemore, E. J. and W. G. Lynard. "Management and Control Technology for Urban Stormwater Pollution," *Journal of the Water Pollution Control Federation*, 54(7):1099–1111 (1982).

Guy, H. P. and G. E. Ferguson. "Sediment in Small Reservoirs Due to Urbanization," *J. Hyd. Div.*, ASCE, 88(HY2):27–37 (1962).

Hunt, P. G. and C. R. Lee. "Land Treatment of Wastewater by Overland Flow for Improved Water Quality," in *Biological Control of Water Pollution*. J. Tourbier et al., eds. Philadelphia: University of Pennsylvania Press, Chapter 18 (1976).

Jong, J. "The Purification of Wastewater with the Aid of Rush or Reed Ponds," in *Biological Control of Water Pollution*. J. Tourbier et al., eds. Philadelphia: University of Pennyslvania Press, Chapter 16 (1976).

Kluesener, J. W. and G. F. Lee. "Nutrient Loading from a Separate Storm Sewer in Madison, Wisconsin," *J. Water Poll. Control Fed.*, 46(5):920–936 (1974).

Lee, L. T. and T. L. Essex. "Urban Headwater Flood Damage Potential," *Journal of Hydraulics Division*, American Society of Civil Engineers, 109(4):519–535 (1983).

Mitchell, W. D. "Effect of Artificial Storage on Peak Flow," U.S. Geological Survey Professional Paper 424-B, pp. 12–14 (1961).

Patrick, R. "The Role of Aquatic Plants in Aquatic Ecosystems," in *Biological Control of Water Pollution*. J. Tourbier, et al., eds. Philadelphia: University of Pennsylvania Press, Chapter 8 (1976).

Robinson, A. R. "Technology for Sediment Control in Urban Areas," *Proceedings, National Conference on Sediment Control*, pp. 49–54 (1969).

Whalen, N. A. "Nonpoint Source Control Guidance Hydrologic Modifications," U.S. Environmental Protection Agency Office of Water Planning and Standards, Washington, DC (1977).

Wischmeier, W. H. and D. D. Smith. "Rainfall Energy and Its Relationship to Soil Loss," *Trans. Am. Geophys. Union*, 39(2):285–291 (1958).

Wischmeier, W. H. and J. V. Mannering. "Relation of Soil Properties to its Erodibility," *Soil Sci. Soc. Am. Proc.*, 33(1):131–137 (1969).

Wischmeier, W. H. et al. "A Soil Erodibility Nomograph for Farmland and Construction Sites," *J. Soil Water Cons.*, Sept./Oct., pp. 189–193 (1971).

Yorke, T. H. "Effects of Sediment Control on Sediment Transport in the Northwest Branch Anacostia River Basin, Montgomery County, Maryland," *J. Res. U.S. Geological Survey*, 3(4):487–494 (1975).

Yorke, T. H. and W. J. Davis. "Effects of Urbanization on Sediment Transport in Bel Pre Creek Basin, Maryland," U.S. Geological Survey Professional Paper 750-B, pp. B218–B223 (1971).

8 | Afterword: Perspectives on the Urban Hydrologic Problem

THIS BOOK HAS presented the various elements of the author's approach to the solution of the urban hydrologic problem (see Figure 1.1). Its major premise is that different scientific inputs are needed to gain an overall understanding of urban hydrology. Accordingly, the urban hydrologic problem would be defined differently by the various scientific disciplines, although there would be general agreement that the problem itself stems from the changes in land surface structure and texture which accompany urbanization. The following are brief summaries of these points of view:

The Engineer: Qualitatively, the surface change accompanying urbanization may be quite clearly described. The urban hydrologic problem for the engineer, therefore, revolves around his or her attempt to quantify the accompanying change in process.

The Architect: The main concern of the architect would be the design of the city itself, i.e., the proximity of the buildings to each other or the lack of sufficient flora for aesthetic and climatic tempering purposes.

The Urban Geographer: The geographer would view the urban hydrologic problem from a historical perspective, and state that the evolution of the physical form of the city was due to historical and economic forces which could not be controlled.

The Planner: The planner would attribute the haphazard evolution of the city to lack of proper planning controls.

The Politician: The politician might ascribe the present problems to the fact that his or her predecessors acted in their own best interest in making decisions, as opposed to acting in the best interests of their constituents.

THE PRIVATE CITIZEN: A CASE IN POINT

Imagine driving to work on a sunny, beautiful morning in Yourtown, U.S.A. Suddenly, you must come to an abrupt halt. You look around and see quite a common sight—the traffic jam. After some time has passed, you have crawled forward enough to see the cause for the slowdown. One lane of the road has been closed, and two lanes of traffic are pushed into one. While passing by the obstacle, you see people with excavating equipment cutting a large ditch in the shoulder of the road, and you are amazed and dismayed by the size of the pipe they are installing, and by the extent of the construction required. You think, "How much will this cost me in taxes? Why is such a large pipe needed? Couldn't this have been avoided?"

WHO IS AT FAULT?

If an effort is made to find out who failed and on whom the blame can be placed for the expense and inconvenience, such an effort would most likely lead in a circle.

It is a simple axiom of planning that if sewer and/or water lines are placed down the side of a street, development is encouraged along that road. When one business locates along the byway, another follows, and then houses and churches. In a relatively short period of time, the sewer and water lines are overtaxed by the strip development, and they must be removed and larger ones installed. Roads follow suit, and soon become congested, and larger roads are needed. If the first businesses fronted on the original road, it becomes a very costly task to purchase or move these establishments to obtain additional rights of way needed for utility improvement. In the end, the resident has to face the ultimate burden of increased taxes.

Where, then, is the blame to be placed—on the city planners? After all, they should have known better. No, most likely not. Perhaps there was no planning development at the time the original sewer line was being proposed. Or, even more likely, planners *had* anticipated the growth described, and had recommended that a larger sewer and water line be installed initially. These recommendations, unfortunately, may have sat on a shelf because the political group capable of enacting them had other, more pressing concerns at the time, or did not find it politically expedient to implement the recommendations.

What about the engineer—should he or she shoulder the blame? Most likely not. He or she is an applied physical scientist who solves problems in the most economic and efficient manner. He or she was presented with certain data, and designed a sewer system which would function adequately according to the prescribed inputs. Moreover, the system did not fail; rather, the input grew beyond the system limits.

Can the political group be blamed? Again, most likely not. Perhaps at the crucial moment they had reelection to consider, and to suggest to the public that a larger sewerage system was needed, causing an increase in taxes, would have been tantamount to political suicide.

Is the public to be blamed, then? Could they have prevented the situation? Again, the answer is likely to be no. The average concerned citizen suffers from two perception problems: (1) he or she sees things for only a short period of time, and, related to this, (2) thinks linearly, i.e., an area is growing at a phenomenal rate, but the casual observer sees only a small amount of change over a short period of time. One cannot properly perceive the rate of increase and the consequences on the locale. He or she believes that if one house were added to a street where five homes already stand, the additional utility needs would be expanded by a linear rate, or 120% of the original rate. This may not be true; in most cases, it is not. The rate may assume a more logarithmic character, i.e., 134%.

Who is to be blamed? In a way, all of these people, in a truly collective sense, since they all possessed certain knowledge and expertise that could have contributed to the solution of the problem. The ultimate solution to urban hydrologic problems lies in a well-coordinated, multi-disciplinary approach along with continuous citizen input. Figure 8.1 illustrates the means of solution. These are:

(1) The engineering definition and analysis of the problem, coordinated with planning
(2) Planning participation in the engineering recommendation process, in order that the anticipatory mechanism is evaluated properly (in the example above, so that larger pipes were added in the planning stages, or so that a zoning ordinance was enacted which would have eliminated the need for larger pipes)

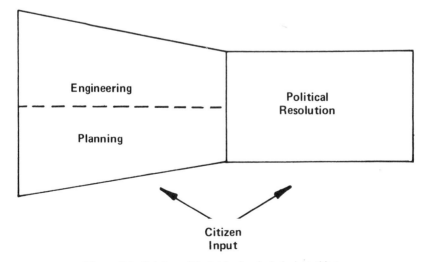

Figure 8.1. Solution of the total urban hydrologic problem.

(3) Coordination between planners and politicians, so that appropriate resolutions are adopted. Public participation is essential to keep all of the professionals involved constantly appraised of the public view and needs.

This outline is concise, simple and straightforward in theory, although difficult to implement in practice. The problem in the past has been myopic thinking, fostered by extreme specialization and professional elitism. We must now focus our efforts to achieve interdisciplinary communication and a truly multidisciplinary approach. The greatness of man's genius is most extraordinarily displayed in the construction and design of his cities; in this genius is also the capability to solve the urban hydrologic problems which result.

Index

aeration, p. 36
aerial photography, pp. 96, 98
approaches, p. 86
  modeling, pp. 124, 125
  probabilistic, pp. 101-104

bacteria, p. 93
bacterial decomposition
  aerobic, p. 38-42
  anaerobic, p. 42
  carbonaceous, p. 40
biotic character, pp. 43-44
BOD (biochemical oxygen demand), pp. 38, 40, 48, 49, 57, 60

channelization
  See urbanized stream channels
COD (chemical oxygen demand), pp. 46, 47, 52, 57, 60
city, p. 1
  anatomy
    See urban form
  block, new design, pp. 218, 129
  break-of-bulk points, p. 3
  growth and urbanization, pp. 3-11
conservatives
  See water quality data collection
constituents, p. 29

depression storage, pp. 131, 157
discharge, pp. 17, 89
dissolved oxygen, pp. 33, 38
Durham, NC (case study), pp. 57-59

erosion
  and water quality, pp. 60-66
  control, pp. 223-227
evaporation, pp. 16, 126
excess rainfall, p. 148

Federal Water Pollution Control Act (PL 92-500)
  See water quality management planning
first flush effect, p. 49
flood
  frequency, p. 102
  recurrence interval, pp. 101, 102, 103
  series, pp. 101, 104, 106

gases, dissolved, p. 92
grass roots planning, p. 205

habitat distribution, pp. 33-37
Hazen and William's formula, p. 137
hydrologic cycle, pp. 15-17
hydrograph, pp. 17-18

impervious area, pp. 20, 22
imperviousness, pp. 20, 21, 24, 124
infiltration, pp. 16, 20, 23, 128, 129, 130, 152, 154

Kutter Formula, p. 137

lagtime, pp. 21, 63, 126
lead, pp. 70, 72
linear system, p. 127
longitudinal profile, pp. 30, 31

239

Index

Manning's formula, p. 136
metals, dissolved, p. 92
Milwaukee, WI (case study), pp. 52-56
modeling approaches, pp. 124-125
models
  "black box," p. 124
  conceptual, of urban water quality modeling, p. 168
  deterministic, pp. 124, 125
  linear, development, pp. 122-124
  linear reservoir, p. 150
  MUNP, p. 160
  parametric, pp. 124, 125
  rainfall-runoff, pp. 127-134
  QUURM, pp. 156, 158
  simulation, p. 146
  stochastic, p. 124
  SWMM, p. 158
Montgomery County, MD (case study), pp. 63-64
multidisciplinary approach, p. xiv

nonconservatives
  See water quality data collection
nonlinear process, p. 132
nonlinear relationships, p. 124
nonlinear responses, p. 140
nonlinearity, p. 126
nonparametric statistics, p. 104
nonpoint source, p. 44
nonstructural control measures, p. 185

organics, p. 92
Ottawa, Ontario (case study), pp. 70, 72
overland flow, pp. 132-134

parametric statistics, p. 104
planner, p. 191
  as a political instrument, p. 198
planning, p. 185
  agencies, pp. 188, 199
  and political decision making, p. 198
  commission, pp. 186, 187
  commission function, p. 188
  commission process, pp. 192, 193
  commission staff, p. 190
  commission structure, p. 187
point source, p. 44
political decision making, p. 198
pollution, pp. 37, 38
pollutograph, pp. 51, 52
precipitation, p. 16
probabilistic approaches, p. 101

Raleigh, NC (case study), pp. 50-52
rational formula or method, pp. 134, 140
reaeration, p. 37
regional planning agency, p. 188
risetime, p. 21
road salt, pp. 66-70
runoff, p. 17
  subsurface, pp. 20, 21
  surface, pp. 20, 22, 23, 128, 134

Scott Run Basin, Fairfax, VA (case study), pp. 64-66
sediment, p. 60
  construction, p. 64
  discharge, p. 65
  sampling, p. 93
  suspended, p. 93
sewage, pp. 37, 38, 39
sewer, p. 134
  design, basic, pp. 135-137
  design, storm, pp. 137-143
  routing, p. 158
sewerage, pp. 44, 134
sheet erosion, prevention, pp. 225-226
statistics
  nonparametric, p. 104
  parametric, p. 104
storage
  depression, pp. 131, 157
  porous pavement, pp. 214-215
  rooftop, pp. 213-214
  surface, p. 17
stormwater
  detention structures, pp. 216-218
  discharge treatment, pp. 220-223
stream
  channels, urbanized, pp. 108-113
  cross section, p. 89
streamflow, pp. 17, 88
street grid, p. 7
street sweeping
  See models, MUNP
structural control measures, p. 213
subwatersheds, p. 149
suburbanization, p. 4
surface
  detention, p. 215
  flow, pp. 157-158
  system, p. 122

unit hydrograph, pp. 20, 21, 22, 127, 143-146

urban anatomy, p. 7
urban flood frequency, p. 102
urban form, p. 6
urban growth, p. 8
urban hydrology, p. xiii
urban surfaces, p. 19
urbanization, pp. 3, 4, 5, 19, 70, 102
urbanized stream channels, p. 108
urbanizing period, p. 86

water quality
  data collection, pp. 91-94
  management planning, p. 201
  modeling, pp. 158-168
  of rivers, pp. 32-33
  sampling, pp. 91-93
  stream modeling, pp. 163-168
  surface runoff modeling, pp. 160-162
water quantity
  modeling, pp. 146-158
  /quality, pp. 29-32
water resources planning, p. 201
watershed, modeling
  natural, pp. 125-127
  urban, pp. 134-146

About the Author

TIMOTHY R. LAZARO is a Professional Hydrologist who earned his Master's degree from the University of Maryland. His thesis examined the effects of urbanization on the streamflow of a watershed.

He was Senior Planner with the Franklin County, Massachusetts, Planning Department and a Physical Planner with the Mount Rogers Planning District Commission, Marion, Virginia. He completed one major land use plan, was project planner for six other plans, and has published several papers related to planning and to urban hydrology. This book represents seven years of graduate and professional research and experience. Parts of the work were published in international technical journals.